ILEANA TOMA

VALERICA MOŞNEGUŢU

ŞTEFANIA CONSTANTINESCU

# INTEGRAL CALCULUS

## AN INTRODUCTION, WITH APPLICATIONS AND EXERCISES

SERIES: MATHEMATICS FOR FUTURE ENGINEERS

VOLUME 4

Printed by CreateSpace

2017

ISBN-13: 978-1548789909

ISBN-10:1548789909

# Preface

This book is addressed to all those who, after finishing the high school, wish a practical initiation in the domain of integral calculus.

This is the fourth volume of the series "Mathematics for future engineering". It has its origins in a course written for the first year students at the Technical University of Constructions of Bucharest.

To provide useful tools for (future) engineers and for specialists, in general, we presented applications of curvilinear, double, surface and triple integrals to mechanics and physics (e.g., centers of mass, moments of inertia, static moments, etc.)

We tried to make the involved mathematics as attractive as possible, by simplifying the presentation without loosing the mathematical rigor of the results. To increase accessibility and to encourage the reader to get a technical know-how about integral calculus, we provided for each newly introduced notion a series of applications and solved problems; each chapter ends by a section containing exercises and problems, each one of these being accompanied by hints and answers.

The sections marked by asterisks can be omitted, as well as some proofs. We introduced them, though, for the sake of a unitary and logical presentation.

The mathematical concepts presented here require only elementary notions, usually learnt in the high school, such as: the notion of integral of functions of one real variable, integration by parts, integrals of rational functions, change of variable in an integral. Actually, the first chapter begins with a recapitulation of these notions, accompanied by numerous examples.

The references contain, along with books, some links with useful sites, which can be helpful for the reader.

We mention that this book is in no way endorsed or sponsored by CreateSpace, Amazon, or their affiliates.

**The Authors**

# CONTENTS

3

# Chapter 1

## THE RIEMANN INTEGRAL

Like differential equations, the integrals are very important in engineering, because by using them one can express many physical and geometrical quantities, which are of interest in technical applications: arc lengths, areas, volumes, centers of mass, moments of inertia, static moments, mechanical work, circulation, etc.

In the following chapters we shall deal with various types of integrals; we shall define, compute and apply curvilinear, double, triple and surface integrals. We shall also establish Green's, Stokes' and flux-divergence formulas, all of them with many applications to the theory of fields, to the strength of materials, as well as to other significant fields of mechanics and physics.

In this chapter, we shall briefly revise the main concepts concerning the notion of integral, already known by the reader from the highschool. We shall define and compute improper (or generalized) integrals, as well as integrals depending on parameters.

# 1.1. RECAPITULATION

## 1.1.1. APPLICATIONS

We consider already known the following integration techniques:

- ♣ integration by parts,
- ♣ changes of variables,
- ♣ computation of integrals with rational integrand.

The table 1.1 contains the most commonly used primitives of elementary functions.

*Table 1.1. Primitves of elementary functions*

| Nr. | $f(x)$ | $\int f(x)dx$ | Nr. | $f(x)$ | $\int f(x)dx$ |
|-----|--------|---------------|-----|--------|---------------|
| 1. | $x^n$ | $\dfrac{x^{n+1}}{n}$ | 7. | $\sin x$ | $-\cos x$ |
| 2. | $e^x$ | $e^x$ | 8. | $\cos x$ | $\sin x$ |
| 3. | $e^{ax}$ | $\dfrac{e^{ax}}{a}$ | 9. | $\dfrac{1}{\cos^2 x}$ | $\tan x$ |
| 4. | $\dfrac{1}{x^2}$ | $-\dfrac{1}{x}$ | 10. | $\dfrac{1}{1+x^2}$ | $\arctan x$ |
| 5. | $xe^x$ | $(x-1)e^x$ | 11. | $x^\alpha$ | $\dfrac{x^{\alpha+1}}{\alpha+1}$ |
| 6. | $\dfrac{1}{x}$ | $\ln|x|$ | 12. | $a^x, a>0$ | $\dfrac{a^x}{\ln a}$ |

We shall recapitulate the following (currently used in applications) integrals, reducible to integrals with rational integrand:

### 1.1.1.1. Integrals of trigonometric functions

They are of the form

$$I \equiv \int_{x_1}^{x_2} R(\sin x, \cos x)\, dx,\qquad (1.1.1)$$

where $R(u,v) = \dfrac{P(u,v)}{Q(u,v)}$, $P,Q$ being polynomials in $u$ and $v$. $R$ is defined on $[x_1, x_2]$ if $Q(\sin x, \cos x) \neq 0,\ x \in [x_1, x_2]$.

### Calculation

1) If $x \neq (2k+1)\pi$, then we use the change of variable

$$t = \tan\frac{x}{2}.\qquad (1.1.2)$$

We have

$$x = 2\arctan t \quad \Rightarrow \quad dx = \frac{2}{1+t^2}\, dt,\qquad (1.1.3)$$

and $\sin x = \dfrac{2t}{1+t^2}$, $\cos x = \dfrac{1-t^2}{1+t}$. The limits of integration are:

$$t_j = \tan\frac{x_j}{2}, j = \overline{1,2}.$$

It results

$$I = \int_{t_1}^{t_2} \underbrace{R\left(\frac{2t}{1+t^2}, \frac{1-t^2}{1+t^2}\right)\cdot\frac{2}{1+t^2}}_{r(t)\to\text{rational function}}\, dt,\qquad (1.1.4)$$

9

which is an integral with rational integrand.

*Example.* Compute the following integral

$$I = \int\limits_0^{\frac{\pi}{2}} \frac{dx}{5+\cos x}.$$

(1.1.5)

**Solution.** We use the change of variable

$$t = \tan\frac{x}{2}, \quad x = 2\arctan t, \quad dx = \frac{2}{1+t^2}dt, \quad \cos x = \frac{1-t^2}{1+t^2}.$$

(1.1.6)

The limits of integration are

$$t_1 = \tan\frac{x_1}{2} = \tan 0 = 0, \quad t_2 = \tan\frac{x_2}{2} = \tan\frac{\pi}{4} = 1.$$

(1.1.7)

Further,

$$I = \int\limits_0^1 \frac{1}{5+\frac{1-t^2}{1+t^2}} \cdot \frac{2}{1+t^2}dt = \int\limits_0^1 \frac{2}{6+4t^2}dt = \frac{1}{3}\int\limits_0^1 \frac{dt}{1+\frac{2}{3}t^2}.$$

(1.1.8)

By using the change of variable

$$u = \sqrt{\frac{2}{3}}t \Rightarrow du = \sqrt{\frac{2}{3}}dt, \quad u_1 = 0, u_2 = \sqrt{\frac{2}{3}}\cdot 1 = \sqrt{\frac{2}{3}},$$

(1.1.9)

we are lead to

$$I = \frac{1}{3}\int\limits_0^{\sqrt{\frac{2}{3}}} \frac{\sqrt{\frac{3}{2}}}{1+u^2}du = \frac{1}{\sqrt{6}}\arctan u \Big|_0^{\sqrt{\frac{2}{3}}}$$

(1.1.10)

$$\Rightarrow \boxed{I = \frac{1}{\sqrt{6}}\arctan\sqrt{\frac{2}{3}}}.$$

10

2) If $R(u,v)$ is **odd** in $u$, i.e., $R(-u,v) = -R(u,v)$, then we make the following change of variable:

$$t = \cos x \Rightarrow dt = -\sin x \cdot dx \Rightarrow t_1 = \cos x_1, \ t_2 = \cos x_2. \quad (1.1.11)$$

As $R$ is odd in $u$, it follows that

$$R(\sin x, \cos x) = \sin x \cdot R_1\left(\sin^2 x, \cos x\right), \quad (1.1.12)$$

with $R_1(u,v)$ rational with respect to the arguments $u$, $v$.

Hence

$$I = \int_{x_1}^{x_2} R(\sin x, \cos x)dx = \int_{x_1}^{x_2} \sin x \cdot R_1\left(\sin^2 x, \cos x\right)dx =$$

$$\quad (1.1.13)$$

$$= \int_{x_1}^{x_2} -R_1\left(1 - \cos^2 x, \cos x\right)d\left(\cos x\right) = \int_{t_1}^{t_2} \underbrace{-R_1\left(1 - t^2, t\right)}_{r(t)\text{rational}}dt.$$

*Example.* Compute the integral

$$I = \int_0^{\frac{\pi}{6}} \frac{\sin^3 x}{1 + \cos x} dx. \quad (1.1.14)$$

**Solution.** The rational function associated to the integrand is odd in $u$

$$R(u,v) = \frac{u^3}{1+v}; \ R(-u,v) = \frac{(-u)^3}{1+v} = -\frac{u^3}{1+v} = -R(u,v). \quad (1.1.15)$$

Given an integrand of this type, we make the change of variable

$$t = \cos x \Rightarrow dt = -\sin x dx. \quad (1.1.16)$$

The limits of integration are

11

$$t_1 = \cos 0 = 1, \quad t_2 = \cos\frac{\pi}{6} = \frac{\sqrt{3}}{2}. \tag{1.1.17}$$

Therefore

$$I = \int\limits_{1}^{\frac{\sqrt{3}}{2}} \frac{1-t^2}{1+t}(-dt) = \int\limits_{\frac{\sqrt{3}}{2}}^{1} (1-t)dt = \left(t - \frac{t^2}{2}\right)\Big|_{\frac{\sqrt{3}}{2}}^{1} = $$

$$\tag{1.1.18}$$

$$1 - \frac{\sqrt{3}}{2} - \frac{1}{2} + \frac{3}{8} \Rightarrow \boxed{I = \frac{7}{8} - \frac{\sqrt{3}}{2}}.$$

### 1.1.1.2 Integrals with irrational integrand

#### A) INTEGRALS OF THE TYPE

$$I \equiv \int\limits_{x_1}^{x_2} R\left(x, \sqrt[n]{\frac{ax+b}{cx+d}}\right) dx, \tag{1.1.19}$$

where $R(u,v) = \dfrac{P(u,v)}{Q(u,v)}$, with $P, Q$ polynomials in $u, v$.

The integral makes sense if:

&clubs;   $Q\left(x, \sqrt[n]{\dfrac{ax+b}{cx+d}}\right) \neq 0, \forall x \in [x_1, x_2]$,

&clubs;   $cx + d \neq 0, \forall x \in [x_1, x_2]$,

&clubs;   $\dfrac{ax+b}{cx+d} \geq 0$, for $n$ even, $\forall x \in [x_1, x_2]$.

We apply the change of variable:

$$t = \sqrt[n]{\frac{ax+b}{cx+d}} \Rightarrow x = \frac{dt^n - b}{a - ct^n} = r(t), \tag{1.1.20}$$

where $r(t)$ is a rational function. It follows that

$$dx = r'(t)dt, \quad t_j = \sqrt[n]{\frac{ax_j + b}{cx_j + d}}, j = \overline{1,2}. \qquad (1.1.21)$$

Hence

$$I = \underbrace{\int_{t_1}^{t_2} R\big(r(t),t\big)r'(t)dt}_{R_1(t)\,\text{rational}}. \qquad (1.1.22)$$

### B) INTEGRALS OF THE TYPE

$$I \equiv$$

$$\int_{x_1}^{x_2} R\left(x, \sqrt[n_1]{\left(\frac{ax+b}{cx+d}\right)^{m_1}}, \sqrt[n_2]{\left(\frac{ax+b}{cx+d}\right)^{m_2}}, \ldots, \sqrt[n_p]{\left(\frac{ax+b}{cx+d}\right)^{m_p}}\right)dx, \qquad (1.1.23)$$

where $n_i, m_i \in \mathfrak{N}, \ i = \overline{1,p}$ and

$$R\big(x, u_1, u_2, \ldots, u_p\big) = \frac{P\big(x, u_1, u_2, \ldots, u_p\big)}{Q\big(x, u_1, u_2, \ldots, u_p\big)}.$$

The integral makes sense if the following conditions are fulfilled on $[x_1, x_2]$:

♣ $Q\left(x, \sqrt[n_1]{\left(\frac{ax+b}{cx+d}\right)^{m_1}}, \sqrt[n_2]{\left(\frac{ax+b}{cx+d}\right)^{m_2}}, \ldots, \sqrt[n_p]{\left(\frac{ax+b}{cx+d}\right)^{m_p}}\right) \neq 0$,

♣ $\dfrac{ax+b}{cx+d} \geq 0$, if at least one of the $n_i, i = \overline{1,p}$, is even,

♣ $cx + d \neq 0$.

Let $n$ be the least common multiple of the positive integers $n_1, n_2, \ldots, n_p$. Then $s_i \equiv \dfrac{n}{n_i} \in \mathfrak{N}$ and we get

$$\sqrt[n_i]{\left(\frac{ax+b}{cx+d}\right)^{m_i}} = \left(\frac{ax+b}{cx+d}\right)^{\frac{m_i}{n_i}} = \left(\frac{ax+b}{cx+d}\right)^{\frac{s_i \cdot m_i}{s_i \cdot n_i}} =$$

$$= \left(\sqrt[n]{\frac{ax+b}{cx+d}}\right)^{m_i s_i}.$$

(1.1.24)

Therefore

$$I =$$

$$\int_{x_1}^{x_2} R\left(x, \underbrace{\sqrt[n]{\left(\frac{ax+b}{cx+d}\right)^{m_1 s_1}}, \sqrt[n]{\left(\frac{ax+b}{cx+d}\right)^{m_2 s_2}}, \ldots, \sqrt[n]{\left(\frac{ax+b}{cx+d}\right)^{m_p s_p}}}_{R_1\left(x, \sqrt[n]{\frac{ax+b}{cx+d}}\right) \quad \Rightarrow \quad case\ A).}\right) dx.$$

(1.1.25)

*Example.* Compute the integral

$$I = \int_{0}^{1} \frac{\sqrt{1-x}-1}{\sqrt[3]{1-x}+1}\, dx.$$

(1.1.26)

**Solution.** As

$$\frac{ax+b}{cx+d} \to 1-x, \quad \begin{cases} p=2,\ n_1=2, n_2=3 \Rightarrow n=6, \\ m_1=m_2=1, \end{cases}$$

(1.1.27)

we make the following change of variable: $t = \sqrt[6]{1-x}$ (the integral makes sense, because the corresponding conditions are fulfilled).

Then

$$I = \int_{1}^{0} \frac{t^3-1}{t^2+1}\, d\left(1-t^6\right) = \int_{0}^{1} \frac{t^3-1}{t^2+1}\cdot 6t^5 dt = 6\int_{0}^{1} \frac{t^8-t^5}{t^2+1}\, dt$$

$$\Rightarrow \boxed{I = -\frac{199}{70}+\frac{3\pi}{2}-3\ln 2}.$$

(1.1.28)

14

## C) INTEGRALS OF THE TYPE

$$I \equiv \int_{x_1}^{x_2} R\left(x, \sqrt{ax^2 + bx + c}\right) dx, \qquad (1.1.29)$$

where $R(u,v) = \dfrac{P(u,v)}{Q(u,v)}$, $P, Q$ polynomials.

The integral makes sense if the following conditions are satisfied on $[x_1, x_2]$:

♣ $Q\left(x, \sqrt{ax^2 + bx + c}\right) \neq 0$,

♣ $ax^2 + bx + c \geq 0$.

We consider the following cases:

**i)** $a > 0$; then we make the change of variable

$$\sqrt{ax^2 + bx + c} = \sqrt{a}\,x + t,$$

which implies

$$x = \frac{t^2 - c}{b - 2\sqrt{a}\,t} = r(t), \quad dx = r'(t)dt,$$

$$t_j = \sqrt{ax_j^2 + bx_j + c}, j = \overline{1,2}. \qquad (1.1.30)$$

Hence,

$$I = \int_{t_1}^{t_2} \underbrace{R\left(r(t), \sqrt{a}\,r(t) + t\right)}_{R_1(t)} r'(t)\, dt. \qquad (1.1.31)$$

*Example.* Compute the following integral:

$$I = \int_0^1 \frac{dx}{\sqrt{x^2 + x + 4}}. \qquad (1.1.32)$$

15

**Solution.** The integral makes sense, because the above mentioned conditions are fulfilled. In this case, $a > 0$, therefore we make the change of variable

$$\sqrt{x^2 + x + 4} = x + t \quad \Rightarrow \quad x^2 + x + 4 = x^2 + 2tx + t^2, \qquad (1.1.33)$$

and it follows that

$$x = \frac{t^2 - 4}{1 - 2t}, \quad dx = \frac{2t(1 - 2t) - (-2)(t^2 - 4)}{(1 - 2t)^2} dt. \qquad (1.1.34)$$

Hence $dx = \dfrac{2(-t^2 + t - 4)}{(1 - 2t)^2} dt$, the limits of integration being $t_1 = 2$, $t_2 = \sqrt{6} - 1$. It results

$$I = \int_{2}^{\sqrt{6}-1} \frac{1}{\dfrac{t^2 - 4}{1 - 2t} + t} \cdot \frac{2(-t^2 + t - 4)}{(1 - 2t)^2} dt =$$

$$\qquad (1.1.35)$$

$$= \int_{2}^{\sqrt{6}-1} \frac{2}{1 - 2t} = -\ln|2t - 1|\Big|_{2}^{\sqrt{6}-1},$$

and, finally,

$$\boxed{I = \ln\frac{2\sqrt{6} + 3}{5}}. \qquad (1.1.36)$$

*ii)* $a < 0$ and the trinomial $ax^2 + bx + c$ has the real roots $\alpha, \beta$. Then the trinomial can be written in the form

$$ax^2 + bx + c = a(x - \alpha)(x - \beta). \qquad (1.1.37)$$

We make the change of variable

16

$$\sqrt{ax^2 + bx + c} = \sqrt{a(x-\alpha)(x-\beta)} = t(x-\alpha), \qquad (1.1.38)$$

thus obtaining

$$a(x-\alpha)(x-\beta) = t^2 (x-\alpha)^2 \Rightarrow x = \frac{a\beta - \alpha t^2}{a - t^2} = r(t). \quad (1.1.39)$$

Further, $dx = r'(t)dt$,

$$\sqrt{a(x-\alpha)(x-\beta)} = t(x-\alpha) = t(r(t)-\alpha).$$

Hence

$$I = \int_{t_1}^{t_2} R\big(r(t), t(r(t)-\alpha)\big)r'(t)dt. \qquad (1.1.40)$$

***iii)*** $a < 0$ and $c > 0$. In this case, we use the following change of variable

$$\sqrt{ax^2 + bx + c} = tx + \sqrt{c}. \qquad (1.1.41)$$

*Example.* Compute the integral

$$I = \int_1^2 \frac{dx}{x\sqrt{-x^2 + 4x + 1}}. \qquad (1.1.42)$$

**Solution.** The integral makes sense, as the corresponding conditions are satisfied.

We also have $a = -1 < 0$ and $c = 1 > 0$, therefore we make the change of variable

$$\sqrt{-x^2 + 4x + 1} = tx + 1. \qquad (1.1.43)$$

We have

$$-x^2 + 4x + 1 = t^2x^2 + 2tx + 1 \quad \Rightarrow \quad x = \frac{4 - 2t}{t^2 + 1},$$

$$dx = \frac{-2\left(t^2 + 1\right) - 2t\left(4 - 2t\right)}{\left(t^2 + 1\right)^2} dt, \quad t_1 = 1, \ t_2 = \frac{\sqrt{5} - 1}{2}. \tag{1.1.44}$$

The integral becomes

$$I = \int_{t_1}^{t_2} \frac{1}{\left(t \cdot \dfrac{4 - 2t}{t^2 + 1} + 1\right) \cdot \dfrac{4 - 2t}{t^2 + 1}} \cdot \frac{2\left(t^2 - 4t - 1\right)}{\left(t^2 + 1\right)^2} dt =$$

$$= \int_{t_1}^{t_2} \frac{\left(t^2 + 1\right)^2}{2(2 - t)\left(-t^2 + 4t + 1\right)} \cdot \frac{2\left(t^2 - 4t - 1\right)}{\left(t^2 + 1\right)^2} dt = \tag{1.1.45}$$

$$= \int_1^{\frac{\sqrt{5} - 1}{2}} \frac{dt}{t - 2} = \ln|t - 2|\Big|_1^{\frac{\sqrt{5} - 1}{2}} = \ln\left(2 - \frac{\sqrt{5} - 1}{2}\right) - \ln 1,$$

i.e.

$$\boxed{I = \ln\frac{5 - \sqrt{5}}{2}}. \tag{1.1.46}$$

*Remark.* This exercise can also be solved by using the method from the point *ii)*.

### 1.1.2. RIEMANN INTEGRABLE FUNCTIONS

#### *1.1.2.1. Definition of the Riemann integral*

Let $f : [a, b] \subset \Re \to \Re$ and consider the partition $\Delta = \{a = x_0, x_1, x_2, \ldots, x_n = b\}$, with $x_i < x_j$ if $i < j$.

*The norm* of $\Delta$ is defined as

$$v(\Delta) = \max\left\{|x_{i+1} - x_i|, \, i = \overline{0, n-1}\right\}. \qquad (1.1.47)$$

Let $\alpha_i \in [x_i, x_{i+1}], \quad i = \overline{0, n-1}$.

**Definition 1.1.** The sum

$$\sigma_\Delta(f, \alpha_i) \equiv \sum_{i=0}^{n-1} f(\alpha_i)(x_{i+1} - x_i) \qquad (1.1.48)$$

is called *the Riemann sum* corresponding to the partition $\Delta$ and to the intermediate points $\alpha_i$.

### GRAPHICAL INTERPRETATION OF THE RIEMANN SUM

If the function $f$ is positive, then the Riemann sum $\sigma_\Delta(f, \alpha_i)$ is the sum of areas of rectangles of width $|x_{i+1} - x_i|$ and of height $f(\alpha_i)$, $i = \overline{0, n-1}$. Therefore, $\sigma_\Delta(f, \alpha_i)$ approximates the area of the plane set, also known as the subgraph of $f$, limited by the $Ox$ axis, by the graph of the function $f$ and by the straight lines parallel to the $Oy$ axis, passing through the points of coordinates $(a, 0)$ and $(b, 0)$ accordingly ( figure 1.1).

**Definition 1.2.** The function $f$ is called *Riemann integrable* on $[a, b]$ if there exists $I \in \mathfrak{R}$ with the property that for any $\varepsilon > 0$ one can find $\eta(\varepsilon) > 0$ such that $\left|\sigma_\Delta(f, \alpha_i) - I\right| < \varepsilon$, for *any partition* $\Delta$ of norm $v(\Delta) < \eta(\varepsilon)$ and for *any choice* of the intermediate points $\alpha_i, i = \overline{0, n-1}$.

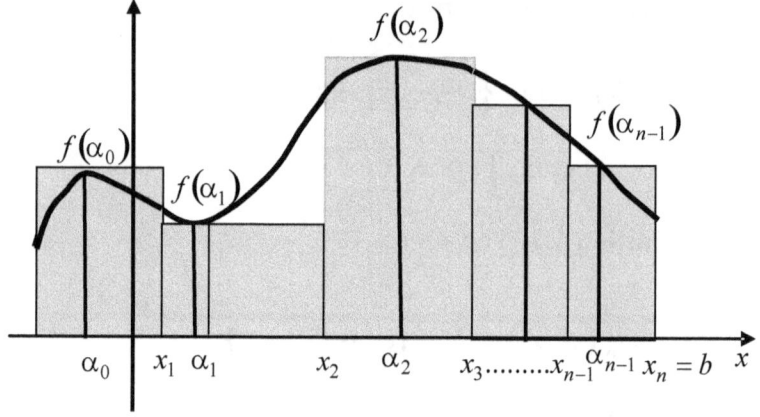

*Figure 1. 1. The Riemann sum as an approximant of the area of a plane set*

In this case, $I$ is **the integral of $f$ from $a$ to $b$** and we write

$$I = \int_a^b f(x)\mathrm{d}x. \tag{1.1.49}$$

### *1.1.2.2. Classes of Riemann integrable functions*

**FUNCTIONS OF BOUNDED VARIATION**

Let $f:[a,b] \subset \Re \to \Re$ (bounded) and $\Delta$ a partition of $[a,b]$.

**Definition 1.3.**

- The **variation** of $f$ with respect to $\Delta$ is the non-negative number

$$V_\Delta(f) = \sum_{i=1}^{n-1} |f(x_{i+1}) - f(x_i)|. \tag{1.1.50}$$

- *f* is called a ***function of bounded variation*** on $[a,b]$ if there exists $M > 0\,(M < \infty)$ such that $|V(f)| < M$, for any partition $\Delta$ of $[a,b]$.

- The ***total variation*** of $f$ on $[a,b]$ is defined as

$$\overset{b}{\underset{a}{V}}(f) = \sup_{\Delta} V_{\Delta}(f). \qquad (1.1.51)$$

**CLASSES OF FUNCTIONS OF BOUNDED VARIATION:**

- monotone functions,
- Lipschitz-ian functions,
- differentiable functions with bounded derivative on $[a,b]$.

## PROPERTIES:

***a)*** If $f:[a,b] \subset \Re \to \Re$ is a function of bounded variation on both $[a,c]$ and $[c,b]$, then $f$ is a function of bounded variation on $[a,b]$ and

$$\overset{b}{\underset{a}{V}}(f) = \overset{c}{\underset{a}{V}}(f) + \overset{b}{\underset{c}{V}}(f). \qquad (1.1.52)$$

***b)* Theorem 1.1.** (***JORDAN'S STRUCTURAL THEOREM***). *Let* $f:[a,b] \subset \Re \to \Re$ *be a function of bounded variation. Then there exists* $\varphi, \psi:[a,b] \to \Re$, *both of them monotonically increasing, such that* $f(x) = \varphi(x) - \psi(x)$, $x \in [a,b]$.

**\* Proof.** We define $\varphi$ as follows:

$$\varphi(x) = \overset{x}{\underset{a}{V}}(x). \tag{1.1.53}$$

The function $\varphi$ is monotonically increasing. Indeed, for $x', x'' \in [a,b]$ with $x' < x''$, we have, obviously, $[a,x'] \subset [a,x'']$. It follows that $\overset{x'}{\underset{a}{V}}(f) < \overset{x''}{\underset{a}{V}}$, which means that $\varphi(x') < \varphi(x'')$.

Now we define $\psi$ by using $\varphi$:

$$\psi(x) = \varphi(x) - f(x). \tag{1.1.54}$$

The function $\psi$ is monotonically increasing. Indeed, take $x', x'' \in [a,b]$, with $x' < x''$, then

$$\psi(x'') - \psi(x') = \underbrace{\left[\varphi(x'') - \varphi(x')\right]}_{\overset{x''}{\underset{x'}{V}}(f)} - \left[f(x'') - f(x')\right] =$$

$$= \overset{x''}{\underset{x'}{V}}(f) - \left[f(x') - f(x'')\right]. \tag{1.1.55}$$

But

$$f(x'') - f(x') \le \left|f(x'') - f(x')\right| < \overset{x''}{\underset{x'}{V}}(f), \tag{1.1.56}$$

therefore

$$\psi(x'') - \psi(x') > 0 \quad \Rightarrow \quad \psi(x') < \psi(x''). \tag{1.1.57}$$

*c)* If $f, g$ are functions of bounded variation on $[a,b]$, then $f \cdot g, |f|$ and $\dfrac{1}{f}$ (if $f$ is non-zero on $[a,b]$) are also functions of bounded variation on $[a,b]$.

*d)* **Theorem 1.2.** *(COROLLARY TO JORDAN'S STRUCTURAL THEOREM). If $f$ is a function of bounded variation on $[a,b]$, then $f$ is Riemann integrable on $[a,b]$.*

In conclusion, from the previous considerations, one can emphasize some

**CLASSES OF RIEMANN INTEGRABLE FUNCTIONS:**

- functions of bounded variation,
- continuous functions.

### * 1.1.2.3. Darboux sums

Let $f : [a,b] \to \Re$, bounded on $[a,b]$.

Let $\Delta$ be a partition of $[a,b]$, $\Delta = \{a = x_0, x_1, x_2, \ldots, x_n = b\}$, with $x_i < x_j$, $i < j$, $i, j = \overline{0,n}$, and denote its norm by $v(\Delta) = \max \{ |x_{i+1} - x_i|, i = \overline{0, n-1} \}$ $v(\Delta) = \max \{ |x_{i+1} - x_i|, i = \overline{0, n-1} \}$.

Let $\alpha_i \in [x_{i-1}, x_i]$, $i = \overline{1,n}$. We consider the Riemann sum

$$\sigma_\Delta (f, \alpha_i) = \sum_{i=0}^{n} f(\alpha_i)(x_i - x_{i-1}).$$ (1.1.58)

For $\Delta$ we shall define the following Darboux sums:

**1. *the upper Darboux sum*:**

$$S_\Delta (f) = \sum_{i=1}^{n} M_i (x_i - x_{i-1}), \quad M_i = \sup_{x \in [x_{i-1}, x_i]} f(x),$$ (1.1.59)

and

**2. *the lower Darboux sum*:**

$$s_\Delta(f) = \sum_{i=1}^{n} m_i (x_i - x_{i-1}), \quad m_i = \inf_{x \in [x_{i-1}, x_i]} f(x). \qquad (1.1.60)$$

We get, for the same partition $\Delta$,

$$s_\Delta(f) \le \sigma_\Delta(f, \alpha_i) \le S_\Delta(f), \qquad (1.1.61)$$

for any choice of the intermediate points $\alpha_i$ in the Riemann sum. For a given partition, the sums $s$ and $S$ are, accordingly, the upper and the lower bounds of the Riemann sums.

The Darboux sums also have the following simple properties:

- If we add new points to a partition, then the lower Darboux sum can only increase, and the upper one can only decrease.
- Any lower Darboux sum does not exceed an upper sum, even if it corresponds to another partition of the interval.

From the above-mentioned properties, it follows that the set $\{s\}$ of the lower Darboux sums is bounded up by any upper Darboux sum and it has a finite upper bound:

$$\underline{I} = \sup_\Delta s_\Delta(f), \qquad (1.1.62)$$

and, similarly, the set $\{S\}$ of upper Darboux sums is bounded below by any lower Darboux sum and it has a finite lower bound

$$\overline{I} = \inf_\Delta S_\Delta(f). \qquad (1.1.63)$$

Obviously,

$$\underline{I} \le \overline{I}, \tag{1.1.64}$$

and

$$s_\Delta(f) \le \underline{I} \le \overline{I} \le S_\Delta(f), \tag{1.1.65}$$

for any lower and upper Darboux sum.

We can prove

**Theorem 1.3.** (*DARBOUX'S CRITERION OF INTEGRABILITY*). *The function* $f:[a,b] \subset \Re \to \Re$ *is integrable on* $[a,b]$ *if and only if for any* $\varepsilon > 0$ *one can find* $\eta = \eta(\varepsilon) > 0$ *such that*

$$S_\Delta(f) - s_\Delta(f) < \varepsilon, \tag{1.1.66}$$

*for any partition* $\Delta$ *of norm smaller than* $\eta$ *(i.e.,* $v(\Delta) < \eta$).

\* **Proof.** NECESSITY $\Rightarrow$ If $f$ is Riemann integrable on $[a,b]$, then there exists a real number $I \in \Re$ with the property that for any $\varepsilon > 0$ one can find $\eta = \eta(\varepsilon)$ such that

$$\left| \sigma_\Delta(f, \alpha_i) - I \right| < \frac{\varepsilon}{2}, \tag{1.1.67}$$

for any partition $\Delta$ with $v(\Delta) < \eta$ and for any choice of the intermediate points $\alpha_i \in [x_{i-1}, x_i]$. It follows that

$$I - \frac{\varepsilon}{2} < \sigma_\Delta(f, \alpha_i) < I + \frac{\varepsilon}{2}. \tag{1.1.68}$$

The Darboux sums also satisfy the relation

$$I - \frac{\varepsilon}{2} < s_\Delta (f) \le S_\Delta (f) < I + \frac{\varepsilon}{2}, \qquad (1.1.69)$$

as it follows from the above-mentioned properties. We immediately conclude that

$$S_\Delta (f) - s_\Delta (f) < \varepsilon. \qquad (1.1.70)$$

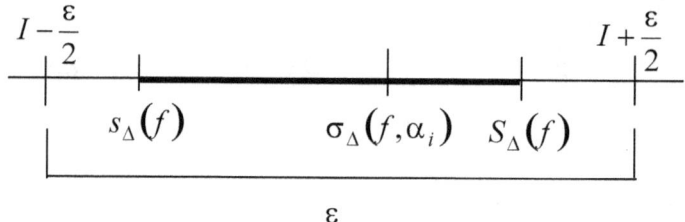

*Figure 1. 2. The distribution of the Darboux sums*

**SUFFICIENCY** $\Leftarrow$ Conversely, we assume that <u>for any</u> $\varepsilon > 0$ <u>there exists a</u> $\eta = \eta(\varepsilon) > 0$ such that

$$S_\Delta (f) - s_\Delta (f) < \varepsilon, \qquad (1.1.71)$$

for <u>any partition</u> $\Delta$ <u>whose norm satisfies the inequality</u> $v(\Delta) < \eta$.

Let $\overline{I} = \inf_\Delta S_\Delta (f), \quad \underline{I} = \sup_\Delta s_\Delta (f)$. We have

$$s_\Delta (f) \le \underline{I} \le \overline{I} \le S_\Delta (f), \qquad (1.1.72)$$

therefore

$$0 \le \overline{I} - \underline{I} \le S_\Delta (f) - s_\Delta (f) \le \varepsilon \quad \Rightarrow \quad \underline{I} = \overline{I} = I. \qquad (1.1.73)$$

The common value $I$ is, actually, the value of the integral. Indeed, let $\sigma_\Delta (f, \alpha_i)$ be a Riemann sum for $\Delta$. From the last relation it follows that $I \in \left[ s_\Delta (f), S_\Delta (f) \right]$ and, as

$\sigma_\Delta(f, \alpha_i) \in \left[ s_\Delta(f), S_\Delta(f) \right]$ for any choice of the intermediate points $\alpha_i$, we obtain

$$\left| \sigma_\Delta(f, \alpha_i) - I \right| < S_\Delta(f) - s_\Delta(f) < \varepsilon.$$

From the underlined relations, it follows that $f$ is Riemann integrable on $[a, b]$, according to the definition. □

Using Darboux's criterion, we can prove

**Theorem 1.4.** *If $f$ is defined and continuous on the real interval $[a, b]$, then $f$ is integrable on $[a, b]$.*

\* **Proof.** Let $\Delta$ be a partition of the interval $[a, b]$. We set up the corresponding Darboux sums

$$S_\Delta(f) = \sum_{i=1}^{n} M_i (x_i - x_{i-1}), \quad M_i = \sup_{x \in [x_{i-1}, x_i]} f(x),$$

$$\tag{1.1.74}$$

$$s_\Delta(f) = \sum_{i=1}^{n} m_i (x_i - x_{i-1}), \quad m_i = \inf_{x \in [x_{i-1}, x_i]} f(x).$$

As $f \in C^0 \left( [a, b] \right)$, it follows that $f$ is bounded and attains its bounds on each interval of the partition, hence

➢ there exists $\overline{\alpha}_i \in [x_{i-1}, x_i]$ such that $M_i = f(\overline{\alpha}_i)$,

➢ there exists $\underline{\alpha}_i \in [x_{i-1}, x_i]$ such that $m_i = f(\underline{\alpha}_i)$.

Now, we majorize the difference:

$$S_\Delta(f) - s_\Delta(f) = \sum_{i=1}^{n} (M_i - m_i)(x_i - x_{i-1}) =$$

$$= \sum_{i=1}^{n} \left[ f(\overline{\alpha}_i) - f(\underline{\alpha}_i) \right](x_i - x_{i-1}) \qquad (1.1.75)$$

Let us take $\varepsilon > 0$. As $f$ is continuous on the closed interval $[a,b]$, it follows that $f$ is also uniformly continuous on $[a,b]$. Hence, one can find $\eta = \eta(\varepsilon)$ such that:

$$\left| f(x') - f(x'') \right| < \varepsilon, \quad x', x'' \in [a,b], \quad \left| x' - x'' \right| < \eta. \qquad (1.1.76)$$

We choose the partition $\Delta$ such that its norm be lesser that $\eta$ (i.e., $v(\Delta) < \eta$). Then

$$\left| f(\overline{\alpha}_i) - f(\underline{\alpha}_i) \right| < \varepsilon. \qquad (1.1.77)$$

This leads to the inequality

$$S_\Delta(f) - s_\Delta(f) < \varepsilon(b - a), \qquad (1.1.78)$$

for any $\Delta$ with $v(\Delta) < \eta$.

The underlined relations imply the integrability of $f$ on $[a,b]$, according to Darboux's criterion. ◻

## 1.2. IMPROPER (GENERALIZED) INTEGRALS

So far, we tackled integrals of bounded functions on finite intervals. However, there are physical models involving the computation of the so-called **generalized** or **improper integrals**. These can be

28

***A. integrals on unbounded intervals or/and***

***B. integrals of unbounded functions.***

We shall study them by turns.

## 1.2.1. INTEGRALS ON UNBOUNDED INTERVALS

**Definition.** Let $f:[a,+\infty)\to\Re$ be integrable on any closed and bounded interval $[a,b]\subset[a,+\infty)$. If $\lim\limits_{b\to\infty}\int\limits_a^b f(x)\mathrm{d}x$ exists and is finite, then we say that $\int\limits_a^\infty f(x)\mathrm{d}x$ ***is convergent*** and

$$\int\limits_a^\infty f(x)\mathrm{d}x = \lim\limits_{b\to\infty}\int\limits_a^b f(x)\mathrm{d}x. \qquad (1.2.1)$$

In the negative, we say that $\int\limits_a^\infty f(x)\mathrm{d}x$ makes no sense.

## CONVERGENCE CRITERIONS

**I. Theorem 1.5.** (***THE CAUCHY INTEGRAL CRITERION***). *Let* $f:[1,+\infty)\to\Re_+$ *be a continuous and monotonically decreasing function with* $\lim\limits_{n\to\infty}f(x)=0$. *Then* $\int\limits_1^\infty f(x)\mathrm{d}x$ *is convergent if and only if the series* $\sum\limits_{j=1}^\infty f(j)$ *is convergent.*

**\* Proof.** We have

$$\int_1^\infty f(x)dx =$$

$$= \int_1^2 f(x)dx + \int_2^3 f(x)dx + \ldots + \int_j^{j+1} f(x)dx + \ldots =$$

(1.2.2)

$$= \sum_{j=1}^\infty \underbrace{\int_j^{j+1} f(x)dx}_{u_j}.$$

Therefore $\int_1^\infty f(x)dx = \sum_{j=1}^\infty u_j$ is written in the form of a series.

Let $F$ be the primitive of $f$, i.e., $F'(x) = f(x)$.

We have

$$u_j = \int_j^{j+1} f(x)dx = F(j+1) - F(j),$$

(1.2.3)

but, according to Lagrange's theorem,

$$F(j+1) - F(j) = F'(\alpha_j)(j+1-j) = f(\alpha_j),$$

(1.2.4)

where $j < \alpha_j < j+1$. As $f$ is decreasing, it results

$$f(j+1) \le f(\alpha_j) \le f(j),$$

(1.2.5)

and, by virtue of the theorem of comparison for series with positive terms, it follows that $\sum_{j=1}^\infty f(\alpha_j)$ is convergent if and only if

$\sum_{j=1}^\infty f(j)$ is convergent.

*Example.* (***THE EULER-POISSON INTEGRAL***) Compute the following integral, known as the Euler-Poisson integral:

$$I = \int_0^\infty e^{-x^2} dx. \tag{1.2.6}$$

**Solution.** We split the integral in two parts:

$$I = \underbrace{\int_0^1 e^{-x^2} dx}_{\text{integral on a bounded interval}} + \int_1^\infty e^{-x^2} dx. \tag{1.2.7}$$

The first integral is easily computed. Let us focus on the second.

The integrand $f(x) = e^{-x^2}$ satisfies the conditions of the theorem 1.1, because

- $f(x) = e^{-x^2}$ is monotonically decreasing,

- $\lim_{x \to \infty} e^{-x^2} = 0$,

- $f(x) = e^{-x^2}$ is a continuous function.

Then, according to the root criterion, $\sum_{n=1}^\infty e^{-n^2}$ is convergent, whence it follows that $I$ is convergent.

**II.** Consider the integral

$$\int_a^\infty \frac{1}{x^p} dx, \qquad p > 0,\ a > 0. \tag{1.2.8}$$

Obviously, $\int_a^b \frac{1}{x^p} dx$ is integrable on any interval $[a, b]$.

For $p \neq 1$, we have

$$I_b \equiv \int_a^b \frac{1}{x^p} dx = \frac{x^{-p+1}}{-p+1}\bigg|_a^b = \frac{1}{1-p}\left(b^{1-p} - a^{1-p}\right). \qquad (1.2.9)$$

Also, $\lim_{b \to \infty} I_b = \begin{cases} \dfrac{a^{1-p}}{p-1}, & \text{for } p > 1, \\ \infty, & \text{for } p < 1. \end{cases}$

If $p = 1$, then $\int_a^b \frac{1}{x} dx = \ln\frac{b}{a}$, and $\lim_{b \to \infty} I_b = \lim_{b \to \infty} \ln\frac{b}{a} = \infty$.

Therefore $\int_a^\infty \frac{1}{x^p} dx$ is

♣   convergent for $p > 1$,

♣   divergent for $p \leq 1$.

We can use this particular case to prove the following convergence criterion for integrals on infinite intervals:

**Theorem 1.6.** *Let* $f : [a, +\infty) \to \Re_+, a > 0$, $f$ *integrable on any* $[a, b] \subset [a, +\infty)$. *Under these hypotheses,*

*i) if there exists* $\alpha > 1$ *such that* $\lim_{x \to \infty} x^\alpha f(x) = q < \infty$,

*then* $\int_a^\infty f(x) dx$ *converges;*

*ii)* *if* *there* *exists* $\alpha \leq 1$ *such* *that*

$\lim_{x \to \infty} x^\alpha f(x) = q \neq 0, q < \infty$, *then* $\int_a^\infty f(x) dx$ *diverges.*

**Proof.**

*i)* From the hypothesis, it follows that one can find $M > 0$ such that

$$f(x) < Mx^{-\alpha} = \frac{M}{x^{\alpha}}, \qquad x \in [a, +\infty), \qquad \text{for } \alpha > 1, \qquad (1.2.10)$$

therefore

$$0 < \int_a^b f(x)\,dx < M \int_a^b \frac{1}{x^{\alpha}}\,dx. \qquad (1.2.11)$$

If we take the limit for $b \to \infty$, as $M \int_a^{\infty} \frac{1}{x^{\alpha}}\,dx$ converges, it

results that $\int_a^{\infty} f(x)\,dx$ also converges.

*ii)* The hypothesis implies that it exists a constant $K > 0$ such that

$$f(x) > \frac{K}{x^{\alpha}}, \qquad (1.2.12)$$

whence it follows that

$$\int_a^b f(x)\,dx > K \int_a^b \frac{dx}{x^{\alpha}}, \qquad \text{for } \alpha \leq 1. \qquad (1.2.13)$$

Taking the limit for $b \to \infty$, we infer that $K \int_a^b \frac{dx}{x^{\alpha}} \to \infty$,

which means that $\int_a^b f(x)\,dx$ diverges. $\square$

*Examples.*

*1)* $\int_a^\infty P(x)e^{-q^2x}dx$, $a > 0$, converges for any polynomial of an arbitrary degree. Indeed, the number of zeros of $P$ is finite and they are situated at finite distances; obviously, we can consider $a$ to be greater than their upper bound. In the negative, we divide the interval in two parts: the interval $[a, b]$, which contains all the zeros of $P$ placed on the right hand side of $a$, and the infinite interval $[b, \infty)$, which contains no zeros of $P$ by construction. Taking this into account, we can consider that $P$ has a constant sign on $[a, \infty)$ and we can apply the theorem 1.6.

*2)* Consider the integral $\int_a^\infty \frac{P(x)}{Q(x)}dx$, where $P, Q$ are polynomials of degree $m$ and $n$ respectively. Suppose that $P(x) \neq 0$, $Q(x) \neq 0$, $x \geq a$, therefore their ratio has a constant sign on $[a, \infty)$; otherwise, we proceed as previously. Let $a_m, b_n$ be the coefficients of the terms of maximum degree in each polynomial. We compute

$$\lim_{x\to\infty} x^\alpha \cdot \frac{P(x)}{Q(x)} = \lim_{x\to\infty} x^\alpha \cdot \frac{a_m}{b_n} \cdot x^{m-n} = \lim_{x\to\infty} \frac{a_m}{b_n} \cdot x^{\alpha+m-n}. \qquad (1.2.14)$$

The limit is finite if $\alpha = n - m$. However, the theorem 1.6 ensures the convergence of the integral for $\alpha > 1$, therefore $n - m > 1$ or $n - m \geq 2$.

In conclusion, $\displaystyle\int_{a}^{\infty}\frac{P(x)}{Q(x)}dx$ *converges if* $n - m \geq 2$ (the degree of the denominator exceeds the degree of the numerator with two units, at least).

Let $f : [a, +\infty) \to \Re$. Inspired by Cauchy's integral criterion (theorem 1.5), we can write the integral of $f$ on the interval $[a, +\infty)$ in the form of a series.

To do this, we consider the partition $\Delta = \{a, a+1, a+2, a+3, \ldots, a+n, \ldots\}$ of $[a, +\infty)$. Then, we can (formally) write

$$\int_{a}^{\infty} f(x)dx =$$

$$\int_{a}^{a+1} f(x)dx + \int_{a+1}^{a+2} f(x)dx + \ldots + \int_{a+n-1}^{a+n} f(x)dx + \ldots \quad (1.2.15)$$

Using the notation $\displaystyle p_n = \int_{a+n-1}^{a+n} f(x)\,dx$, we get

$$\int_{a}^{\infty} f(x)dx = p_1 + p_2 + \ldots + p_n + \ldots = \sum_{n=1}^{\infty} p_n. \quad (1.2.16)$$

From the above, we conclude that $\displaystyle\int_{a}^{\infty} f(x)dx$ converges if and only if the numerical series $\displaystyle\sum_{n=1}^{\infty} p_n$ converges; hence, according to the fundamental theorem for numerical series, if and only if for any $\varepsilon > 0$ there is a rank $N_\varepsilon$, such that

$$\left|R_{n,m}\right| \equiv \left|p_{n+1} + p_{n+2} + \ldots + p_{n+m}\right| < \varepsilon, \qquad n > N_\varepsilon, \ m \in \mathfrak{N}. \quad (1.2.17)$$

In conclusion, one can state another general criterion of convergence for integrals on infinite interval.

**Theorem 1.7.** *Let* $f : [a, +\infty) \to \mathfrak{R}$ *be integrable on* $[a, b] \subset [a, +\infty)$. *Then* $\displaystyle\int_a^\infty f(x)\,dx$ *is convergent if and only if*

$$\lim_{A, B \to \infty} \int_A^B f(x)\,dx = 0.$$

\* **Proof.** Indeed, from (1.2.16) and (1.2.17) it follows that $\displaystyle\int_a^\infty f(x)\,dx$ converges if and only if for any $\varepsilon > 0$ there exists $N_\varepsilon$ such that

$$\left|R_{n,m}\right| =$$

$$\left| \int_{a+n}^{a+n+1} f(x)\,dx + \int_{a+n+1}^{a+n+2} f(x)\,dx + \ldots + \int_{a+n+m-1}^{a+n+m} f(x)\,dx \right| < \varepsilon, \quad (1.2.18)$$

for any $n > N_\varepsilon$, $m \in \mathfrak{N}$, therefore if and only if

$$\left| \int_{a+n}^{a+n+m} f(x)\,dx \right| < \varepsilon, \qquad n > N_\varepsilon, \ m \in \mathfrak{N}. \ \blacksquare \qquad (1.2.19)$$

**Definition 1.4.** Let $f : [a, +\infty) \to \mathfrak{R}$ be a function changing of sign infinitely many times. We say that $\displaystyle\int_a^\infty f(x)\,dx$ is

***absolutely convergent*** if $\displaystyle\int_a^\infty \left|f(x)\right|\,dx$ is convergent.

As in the case of series, we can prove

**Theorem 1.8.** *If* $\int\limits_a^\infty f(x)\,dx$ *is absolutely convergent, then*

$\int\limits_a^\infty f(x)\,dx$ *is convergent.*

\* **Proof.** By virtue of theorem 1.7, as $\int\limits_a^\infty |f(x)|\,dx$ is convergent, it follows that for any $\varepsilon > 0$ we can find an index $N_\varepsilon$ such that

$$\int\limits_A^B f(x)\,dx < \varepsilon, \qquad A, B > N_\varepsilon. \tag{1.2.20}$$

By the monotonicity property of the Riemann integral, it results that

$$\int\limits_A^B |f(x)|\,dx > \left|\int\limits_A^B f(x)\,dx\right|, \qquad B > A > N_\varepsilon, \tag{1.2.21}$$

because $|f(x)| \geq \pm f(x)$.

We use the theorem 1.7, this time, the converse. $\blacksquare$

*Example.* Study the convergence of the integrals

$$I_1 = \int\limits_0^\infty e^{-p^2 x} \sin qx\,dx, \quad I_2 = \int\limits_0^\infty e^{-p^2 x} \cos qx\,dx \ . \tag{1.2.22}$$

We have

$$\int\limits_0^l |e^{-p^2 x} \sin qx|\,dx \leq \int\limits_0^l e^{-p^2 x}\,dx < \frac{1}{p^2}, \quad l > 0, \tag{1.2.23}$$

therefore $I_1$ is absolutely convergent. According to the theorem 1.8, $I_1$ is convergent. With a similar proof, the integral $I_2$ also converges.

### 1.2.2. INTEGRALS WITH UNBOUNDED INTEGRAND

Generally speaking, they are of the form

$$\int_a^b f(x)\,dx, \quad \lim_{x\to a_+} f(x) = \pm\infty. \qquad (1.2.24)$$

**Definition 1.5.**

*i)* Let $f:(a,b] \subset \Re \to \Re$ be integrable on any closed subinterval $[c,b] \subset (a,b]$; suppose that $\lim_{x\to a_+} f(x) = \infty$. If $\lim_{c\to a_+} \int_c^b f(x)dx$ exists and is finite, then we say that $\int_a^b f(x)dx$ *converges* (or makes sense) and we write

$$\int_a^b f(x)\,dx = \lim_{c\to a_+} \int_c^b f(x)dx. \qquad (1.2.25)$$

Otherwise, we say that $\int_a^b f(x)dx$ *diverges* (or makes no sense).

*ii)* Similarly, if $f:[a,b) \subset \Re \to \Re$ is such that $\lim_{x\to b_-} f(x) = \infty$, $f(x)$ being integrable on any closed subinterval

38

$[a,c] \subset [a,b)$, then, if $\lim\limits_{c \to b_-} \int\limits_a^c f(x)dx$ exists and is finite,

$\int\limits_a^b f(x)dx$ is called **convergent.**

In this case,

$$\int\limits_a^b f(x)\,dx = \lim_{c \to b_-} \int\limits_a^c f(x)dx. \qquad (1.2.26)$$

Otherwise, we say that the integral **diverges** (or makes no sense).

In order to check the convergence of an integral from case *i)*, if $b - a > 1$ we can (formally) write the integral in the form of a series:

$$\int\limits_a^b f(x)\,dx =$$

$$= \underbrace{\int\limits_{a+1}^b f(x)dx}_{\text{makes sense}} + \int\limits_{a+\frac{1}{2}}^{a+1} f(x)dx + \ldots + \underbrace{\int\limits_{a+\frac{1}{n+1}}^{a+\frac{1}{n}} f(x)dx}_{u_n} + \ldots, \qquad (1.2.27)$$

whence

$$\int\limits_a^b f(x)\,dx = \sum_{n=1}^{\infty} u_n, \qquad (1.2.28)$$

therefore the convergence of the integral can be checked up by using the convergence of the series $\sum\limits_{n=1}^{\infty} u_n$.

Using again the fundamental theorem for series, we can prove:

**Theorem 1.9.** *Let* $f : (a, b] \to \Re$, $\lim\limits_{x \to a_+} f(x) = \infty$. *Then*

$\int\limits_a^b f(x)dx$ *is convergent if and only if for any* $\varepsilon > 0$ *we can find*

$\eta_\varepsilon$ *such that*

$$\left| \int\limits_{a+\eta_1}^{a+\eta_2} f(x)dx \right| < \varepsilon, \quad \eta_1 < \eta_2 < \varepsilon. \tag{1.2.29}$$

**The proof** is similar to that from theorem 1.8.∎

*Example.* Study the convergence of the integral

$$I = \int\limits_a^b \frac{dx}{(x-a)^\alpha}. \tag{1.2.30}$$

**Solution.** Let us compute the integral on an interval $[c, b]$, $a < c < b$. We get

$$\int\limits_c^b \frac{dx}{(x-a)^\alpha} = \left. \frac{(x-a)^{-\alpha+1}}{-\alpha+1} \right|_{x=c}^{x=b} = \frac{(b-a)^{1-\alpha}}{1-\alpha} - \frac{(c-a)^{1-\alpha}}{1-\alpha}. \tag{1.2.31}$$

If $c \to a$, we obtain

$$I = \begin{cases} \dfrac{(b-a)^{1-\alpha}}{1-\alpha} & \text{for} \quad \alpha < 1, \\ \infty & \text{for} \quad \alpha > 1. \end{cases} \tag{1.2.32}$$

If $\alpha = 1$, then $I = \int\limits_a^b \frac{1}{x-a}dx = \ln(x-a)\big|_a^b \to \infty$, hence it

diverges.

In conclusion, the integral $I$

➢ converges for $\alpha < 1$,

40

➤ diverges for $\alpha \geq 1$,

according to the definition.

Using this example, we can prove:

**Theorem 1.10.** *Let* $f : (a, b] \to \Re_+$ *be integrable on any interval* $[c, b] \subset (a, b]$, *and such that* $\lim\limits_{x \to a_+} f(x) = \infty$. *Under these conditions,*

i) *if for* $\alpha < 1$ $\lim\limits_{x \to a_+} (x - a)^\alpha f(x) = q < \infty$, *then*

$$I = \int_a^b f(x)\, dx \ converges;$$

ii) *if for* $\alpha \geq 1$ $\lim\limits_{x \to a_+} (x - a)^\alpha f(x) = q \neq 0,\ q < \infty$, *then*

$$I = \int_a^b f(x)\, dx \ diverges.$$

**Proof.**

i) From the hypothesis, it follows that there exists $M > 0$ such that

$$f(x) < \frac{M}{(x - a)^\alpha}, \quad \text{for every } x \in (a, b], \tag{1.2.33}$$

therefore

$$0 < \lim\limits_{c \to a_+} \int_c^b f(x)\, dx \leq M \int_a^b \frac{dx}{(x - a)^\alpha} = \frac{M}{1 - \alpha} \cdot \frac{1}{(b - a)^{\alpha - 1}}. \tag{1.2.34}$$

ii) The hypothesis implies that one can find a constant $K > 0$ such that

$$f(x) > \frac{K}{(x-a)^{\alpha}}, \quad \text{for every } x \in (a,b], \tag{1.2.35}$$

whence we infer

$$\int_{c}^{b} f(x) dx > K \int_{c}^{b} \frac{dx}{(x-a)^{\alpha}}. \tag{1.2.36}$$

As $\lim\limits_{c \to a_{+}} \int_{c}^{b} \frac{dx}{(x-a)^{\alpha}} = \infty$, it is easily seen that

$I = \int\limits_{a}^{b} f(x) dx$ diverges. $\blacksquare$

*Example.* Study the convergence of the integral

$$I = \int\limits_{3}^{7} \frac{dx}{(x-3)^{\frac{2}{3}}}. \tag{1.2.37}$$

**Solution.** We compute

$$\lim_{x \to 3_{+}} (x-3)^{\alpha} (x-3)^{-\frac{2}{3}} = 1, \text{ if } \alpha = \frac{2}{3} < 1.$$

Therefore, by virtue of theorem 1.10, point *i*), *I* converges.

Consider now a function $f : [a,b) \to \mathfrak{R}_{+}$ and such that $\lim\limits_{x \to b_{-}} f(x) = \infty$.

We can prove:

**Theorem 1.11.** *Consider the function $f : [a,b) \to \mathfrak{R}_{+}$, $f$ being integrable on any interval $[a,c] \subset [a,b)$ and such that $\lim\limits_{x \to b_{-}} f(x) = +\infty$. In this case,*

42

**i)** *if for* $\alpha < 1$ $\lim\limits_{x \to b_-} (b-x)^\alpha f(x) = q < \infty$, *then the integral*

$$\int_a^b f(x)dx \quad converges,$$

**ii)** *if for* $\alpha \geq 1$ $\lim\limits_{x \to b_-} (b-x)^\alpha f(x) = q < \infty$, $q \neq 0$, *then the*

*integral* $\int_a^b f(x)dx$ *is divergent.*

**The proof** is similar to the proof in theorem 1.10. ◘

### 1.2.3. EULER'S INTEGRALS

### *1.2.3.1. THE GAMMA FUNCTION,* $\Gamma(p)$

By definition, the Gamma function is

$$\Gamma(p) = \int_0^\infty x^{p-1}e^{-x}dx. \tag{1.2.38}$$

This is an improper integral of type 1.2.1 and also of type 1.2.2, if $p < 1$.

Let us study the convergence of this integral.

**Theorem 1.12.** $\Gamma(p)$ *is convergent for* $p > 0$.

\* **Proof.** We split the integral in two parts:

$$\int_0^\infty x^{p-1}e^{-x}dx = \underbrace{\int_0^a x^{p-1}e^{-x}dx}_{\text{case B } (p<1)} + \underbrace{\int_a^\infty x^{p-1}e^{-x}dx}_{\text{case A}} \equiv J_1 + J_2, \tag{1.2.39}$$

where $a > 0$.

***THE CONVERGENCE OF*** $J_1$: The integrand $f(x) = \dfrac{e^{-x}}{x^{1-p}}$ is

unbounded in $x = 0$, for $p < 1$. We compute

$$\lim_{x \to 0_+} x^{\alpha} \cdot \frac{e^{-x}}{x^{1-p}} = \lim_{x \to 0_+} x^{\alpha - (1-p)} = 1 \quad \text{if} \quad \alpha = 1 - p. \qquad (1.2.40)$$

In order for $J_1$ to be convergent, it is sufficient that $\alpha = 1 - p < 1$, i.e. $p > 0$.

***THE CONVERGENCE OF*** $J_2$: The integrand $f(x) = e^{-x} \cdot x^{p-1}$ has a maximum point at $x = p - 1$. Indeed,

$$f'(x) = (p-1)x^{p-2}e^{-x} + (-1)x^{p-1}e^{-x},$$

which yields $f'(x) = x^{p-2}e^{-x}(p - 1 - x)$. Hence,

➢ if $p - 1 > 0$, we put $a = p - 1$. Then $f(x)$ is monotonically decreasing on $[a, \infty)$, $\lim_{x \to \infty} f(x) = 0$, and thus $J_2$ ***converges***;

➢ if $p - 1 < 0$, the integrand is monotonically decreasing for $\forall a$, on $[a, \infty)$ and it also follows that $J_2$ ***converges***. ◻

**RECURRENCE FORMULAS FOR** $\Gamma(p)$

*a)* $\Gamma(p) = (p-1)\Gamma(p-1)$ for $p > 1$.

**Proof.** We integrate by parts:

$$\Gamma(p) = \int_0^\infty x^{p-1} e^{-x} dx = \left(-x^{p-1} e^{-x}\right)\Big|_{x=0}^{x=\infty} + (p-1)\underbrace{\int_0^\infty x^{p-2} e^{-x} dx}_{\Gamma(p-1)} = \tag{1.2.41}$$

$$= (p-1)\Gamma(p-1).$$

**b)** $\Gamma(n) = (n-1)!, \ n \in \mathcal{N}$.

**Proof.** For $n = 1$,

$$\Gamma(1) = \int_0^\infty x^{1-1} e^{-x} dx = \left(-e^{-x}\right)\Big|_{x=0}^{x=\infty} = 1. \tag{1.2.42}$$

For $n > 1$, by using *a)* repeatedly, we obtain

$$\Gamma(n) = (n-1)\Gamma(n-1) = (n-1)(n-2)\Gamma(n-2) = \ldots =$$
$$= (n-1)(n-2)\ldots 2 \cdot 1 \cdot \Gamma(1) = (n-1)!. \tag{1.2.43}$$

*c)* The combinations of $n$ things $k$ at a time can also be expressed by using $\Gamma$ :

$$C_n^k = \frac{n!}{k!(n-k)!} = \frac{\Gamma(n+1)}{\Gamma(n-k+1)\Gamma(k+1)}. \tag{1.2.44}$$

### 1.2.3.2. THE BETA FUNCTION, $\beta(p,q)$

The **Beta function** is, by definition,

$$\beta(p,q) = \int_0^1 x^{p-1} (1-x)^{q-1} dx. \tag{1.2.45}$$

For $p < 1$ and/or $q < 1$, the integral has an unbounded integrand, hence this is the case 1.2.2. As in the case of $\Gamma$, we can prove:

**Theorem 1.13.** *The function* $\beta(p,q)$ *is convergent for* $p > 0, q > 0$.

* **Proof.** We split the integral in two parts:

$$\beta(p,q) = \int_0^c x^{p-1}(1-x)^{q-1}\,dx + \int_c^1 x^{p-1}(1-x)^{q-1}\,dx \equiv J_1 + J_2, (1.2.46)$$

where $0 < c < 1$.

***THE CONVERGENCE OF*** $J_1$: $f(x) = x^{p-1}(1-x)^{q-1}$ is bounded in $x = 0$ for $p > 1$ and unbounded for $p < 1$. We use the theorem 1.10, thus getting

$$\lim_{x \to 0_+} x^\alpha f(x) = \lim_{x \to 0_+} x^{\alpha+p-1}(1-x)^{q-1} = 1 \quad \text{for} \quad \alpha = 1 - p. \ (1.2.47)$$

The convergence holds true if $\alpha = 1 - p < 1$, hence for $p > 0$. Consequently, $J_1$ converges for $p > 0$ and for any $q$.

***THE CONVERGENCE OF*** $J_2$: $f(x) = x^{p-1}(1-x)^{q-1}$ is unbounded in $x = 1$ for $q < 1$ and bounded for $q > 1$.

We use the theorem 1.11. We compute the limit

$$\lim_{x \to 1_-}(1-x)^\alpha \cdot f(x) = \lim_{x \to 1_-} x^{p-1}(1-x)^{\alpha+q-1} = \lim_{x \to 1_-}(1-x)^{\alpha+q-1}, \ (1.2.48)$$

which is 1 for $\alpha = 1 - q$. According to the theorem 1.11, $J_2$ is convergent if $\alpha < 1$, therefore $J_2$ converges for $q > 0$. $\blacksquare$

**RECURRENCE FORMULAS FOR** $\beta(p,q)$.

*a)* $\beta(p,q) = \dfrac{p-1}{p+q-1}\beta(p-1,q), \ p > 1, q > 0$.

**Proof.** We integrate by parts:

46

$$\beta(p,q) = \int_0^1 x^{p-1}(1-x)^{q-1}\,dx =$$

$$= -\frac{1}{q}\cdot x^{p-1}(1-x)^q\Big|_0^1 + \frac{p-1}{q}\int_0^1 x^{p-2}(1-x)^q\,dx =$$

$$= \frac{p-1}{q}\left[\int_0^1 x^{p-2}(1-x)^{q-1}(1-x)\,dx\right] = \tag{1.2.49}$$

$$= \frac{p-1}{q}\left[\underbrace{\int_0^1 x^{p-2}(1-x)^{q-1}\,dx}_{\beta(p-1,q)} - \underbrace{\int_0^1 x^{p-1}(1-x)^{q-1}\,dx}_{\beta(p,q)}\right].$$

Furthermore,

$$\beta(p,q)\left(1+\frac{p-1}{q}\right) = \frac{p-1}{q}\beta(p-1,q), \tag{1.2.50}$$

whence formula *a)* follows at once.

*b)* $\beta(p,q) = \dfrac{q-1}{p+q-1}\beta(p,q-1),\ p > 0, q > 1$.

**The proof** is similar.

*c)* $\beta(m,n) = \dfrac{(m-1)!(n-1)!}{(m+n-1)!},\ m,n \in \mathcal{N}$.

**Proof.** Starting from

$$\beta(1,1) = \int_0^1 x^{1-1}(1-x)^{1-1}\,dx = 1, \tag{1.2.51}$$

we obtain, step by step, applying repeatedly the two previous formulas:

$$\beta(m,n) = \frac{m-1}{m+n-1}\beta(m-1,n) =$$

$$= \frac{(m-1)(m-2)}{(m+n-1)(m+n-2)}\beta(m-2,n) = \ldots =$$

$$= \ldots \frac{(m-1)!}{(m+n-1)(m+n-2)\ldots(n+1)}\beta(1,n) = \quad (1.2.52)$$

$$= \frac{(m-1)!(n-1)}{(m+n-1)(m+n-2)\ldots(n+1)n}\beta(1,n-1) =$$

$$= \ldots = \frac{(m-1)!(n-1)!}{(m+n-1)!}\beta(1,1),$$

which means **c)**.

**d)** $\beta\left(\dfrac{1}{2},\dfrac{1}{2}\right) = \pi$.

**Proof.** We write the function in the form

$$\beta\left(\frac{1}{2},\frac{1}{2}\right) = \int_0^1 x^{\frac{1}{2}-1}(1-x)^{\frac{1}{2}-1}\,dx = \int_0^1 \frac{dx}{\sqrt{x(1-x)}}, \quad (1.2.53)$$

and we make the change of variable

$$x = \sin^2 t \quad \Rightarrow \quad dx = 2\sin t \cos t\, dt. \quad (1.2.54)$$

The limits of integration are:

$$x_1 = 0 \quad \Rightarrow \quad t_1 = 0,$$
$$x_2 = 1 \quad \Rightarrow \quad t_2 = \frac{\pi}{2}. \quad (1.2.55)$$

It results that

$$\beta\left(\frac{1}{2},\frac{1}{2}\right) = \int_0^{\frac{\pi}{2}} \frac{1}{\sin t \cos t}\cdot 2\sin t \cos t\, dt = 2\cdot\frac{\pi}{2} = \pi. \quad (1.2.56)$$

*e)* $\beta(p,q) = \dfrac{\pi}{\sin \pi p}$ , for $p+q=1$ (without proof).

*f)* $\beta(p,q) = \int\limits_{0}^{\infty} t^{p-1}(1+t)^{-(p+q)}\, dt$ .

**Proof.** We make the change of variable $x = \dfrac{t}{1+t}$,

obtaining, step by step

$$dx = \frac{1 \cdot (1+t) - 1 \cdot t}{(1+t)^2}\, dt = \frac{1}{(1+t)^2}\, dt \,,$$

$$t = \frac{x}{1-x} \quad \Rightarrow \quad \begin{cases} x \to 0 \Rightarrow t = 0 \\ x \to 1_- \Rightarrow t \to \infty \end{cases}, \qquad (1.2.57)$$

$$1 - x = 1 - \frac{t}{1+t} = \frac{1}{1+t} \,.$$

Therefore

$$\beta(p,q) =$$
$$= \int\limits_{0}^{1} x^{p-1}(1-x)^{q-1}\, dx = \int\limits_{0}^{\infty} \frac{t^{p-1}}{(1+t)^{p-1}} \cdot \frac{1}{(1+t)^{q-1}} \cdot \frac{dt}{(1+t)^2} = \qquad (1.2.58)$$
$$= \int\limits_{0}^{\infty} \frac{t^{p-1}}{(1+t)^{p+q}}\, dt,$$

whence we get *f)*.

**RELATIONS BETWEEN THE FUNCTIONS $\Gamma(p)$ AND**

$$\beta(p,q)$$

**Theorem 1.14.** *The following relation holds true*

$$\beta(p,q) = \frac{\Gamma(p) \cdot \Gamma(q)}{\Gamma(p+q)}, \quad p, q > 0. \tag{1.2.59}$$

**\* Proof.** We make the change of variable $x = ty$ $(t > 0)$ in $\Gamma(p)$. It follows that $dx = t\,dy$, and the limits of integration are the same. We have:

$$\Gamma(p) = \int_0^\infty x^{p-1}e^{-x}dx = \int_0^\infty t^{p-1}y^{p-1}e^{-ty}t\,dy. \tag{1.2.60}$$

It follows that

$$\frac{\Gamma(p)}{t^p} = \int_0^\infty y^{p-1}e^{-ty}dy; \tag{1.2.61}$$

analogously,

$$\frac{\Gamma(p)}{t^p} = \int_0^\infty y^{q-1}e^{-ty}dy. \tag{1.2.62}$$

If we replace $t$ with $1+t$ and $p$ with $p+q$ in (1.2.61), we obtain

$$\frac{\Gamma(p+q)}{(1+t)^p} = \int_0^\infty y^{p+q-1}e^{-(t+1)y}dy. \tag{1.2.63}$$

Changing the roles of the parameters $t$ and $y$ in (1.2.61), we get

$$\frac{\Gamma(p)}{y^p} = \int_0^\infty t^{p-1}e^{-ty}dt. \tag{1.2.64}$$

Let us multiply the relation (1.2.63) by $t^{p-1}$ and integrate it on $[0, \infty)$:

$$\Gamma(p+q)\underbrace{\int_0^\infty t^{p-1} \cdot (1+t)^{-(p+q)}\, dt}_{f)} =$$

$$= \int_0^\infty t^{p-1}dt \int_0^\infty y^{p+q-1} \cdot e^{-(t+1)y}dy = \qquad (1.2.65)$$

$$= \int_0^\infty y^{p+q-1} \cdot e^{-y}dy \underbrace{\int_0^\infty t^{p-1} \cdot e^{-ty}dt}_{(1.2.63)} = \int_0^\infty y^{p+q-1} \cdot e^{-y} \cdot \frac{\Gamma(p)}{y^p}dy.$$

It follows that

$$\Gamma(p+q) \cdot \beta(p,q) = \Gamma(p) \cdot \underbrace{\int_0^\infty y^{q-1} \cdot e^{-y}dy}_{\Gamma(q)}. \qquad (1.2.66)$$

## APPLICATIONS

**I.** Compute $\Gamma\left(\dfrac{1}{2}\right)$.

**Solution.** We start by applying formula (1.2.59):

$$\beta\left(\frac{1}{2},\frac{1}{2}\right) = \frac{\Gamma\left(\dfrac{1}{2}\right) \cdot \Gamma\left(\dfrac{1}{2}\right)}{\Gamma\left(\dfrac{1}{2}+\dfrac{1}{2}\right)} = \left[\Gamma\left(\frac{1}{2}\right)\right]^2. \qquad (1.2.67)$$

But $\beta\left(\dfrac{1}{2},\dfrac{1}{2}\right) = \pi$. Therefore

$$\boxed{\Gamma\left(\frac{1}{2}\right) = \sqrt{\pi}}. \qquad (1.2.68)$$

**II.** Compute *the Euler-Poisson integral*

$$I = \int_0^\infty e^{-x^2} dx. \tag{1.2.69}$$

**Solution.** We make the change of variable

$$x = t^{\frac{1}{2}} \quad \Rightarrow \quad dx = \frac{1}{2} t^{-\frac{1}{2}} dt.$$

The limits of integration remain the same; it results

$$I = \frac{1}{2} \int_0^\infty e^{-t} \cdot t^{-\frac{1}{2}} dt = \frac{1}{2} \Gamma\left(\frac{1}{2}\right) = \frac{\sqrt{\pi}}{2}, \tag{1.2.70}$$

therefore

$$\boxed{I = \int_0^\infty e^{-x^2} dx = \frac{\sqrt{\pi}}{2}}. \tag{1.2.71}$$

The Euler-Poisson integral is frequently used in the theory of errors. Considering this integral, we can use **the Fresnel error functions**, very useful in applications:

$$erf = \frac{2}{\sqrt{\pi}} \int_0^t e^{-x^2} dx, \quad erfc = \frac{2}{\sqrt{\pi}} \int_t^\infty e^{-x^2} dx. \tag{1.2.72}$$

We notice that their sum is 1. Indeed,

$$erf + erfc = \frac{2}{\sqrt{\pi}} \int_0^t e^{-x^2} dx + \frac{2}{\sqrt{\pi}} \int_t^\infty e^{-x^2} dx =$$

$$= \frac{2}{\sqrt{\pi}} \left( \int_0^t e^{-x^2} dx + \int_t^\infty e^{-x^2} dx \right) = \tag{1.2.73}$$

$$= \frac{2}{\sqrt{\pi}} \int_0^\infty e^{-x^2} dx = \frac{2}{\sqrt{\pi}} \cdot \frac{\sqrt{\pi}}{2} = 1.$$

## 1.3. INTEGRALS DEPENDING ON PARAMETERS

They are (formally) defined as

$$F(\alpha) \equiv \int_{p(\alpha)}^{q(\alpha)} f(x,\alpha)\,\mathrm{d}x. \qquad (1.3.1)$$

For the consistency of this definition, we assume that

- $f : [a,b] \times I \to \Re$.

- $p, q : I \to \Re$, and $\big[ p(\alpha), q(\alpha) \big] \subseteq [a,b], \forall \alpha \in I$.

We state, without proof,

**Theorem 1.15.** (*TAKING LIMITS UNDER THE INTEGRAL SIGN*) *Let* $f : [a,b] \times I \to \Re$ *be a continuous function on* $[a,b], \forall \alpha \in I$. *Suppose that there exists* $g : [a,b] \to \Re$, *such that* $\lim_{\alpha \to \alpha_0} f(x,\alpha) = g(x)$, $\forall x \in [a,b]$, *uniformly on* $[a,b]$. *Then*

$$\lim_{\alpha \to \alpha_0} \int_a^b f(x,\alpha)\,\mathrm{d}x = \int_a^b \lim_{\alpha \to \alpha_0} f(x,\alpha)\,\mathrm{d}x = \int_a^b g(x)\,\mathrm{d}x. \qquad (1.3.2)$$

**Theorem 1.16.** (*DIFFERENTIATION WITH RESPECT TO THE PARAMETER*). *Let* $f : [a,b] \times I \to \Re$, *f continuous on* $[a,b]$ *(with respect to x). Let* $p, q : I \to [a,b]$. *If*

*i)* $p, q \in C^1(I)$,

*ii)* $\dfrac{\partial f}{\partial \alpha}$ *is continuous on* $I$ *(with respect to* $\alpha$*),*

*then* $F(\alpha) = \displaystyle\int_{p(\alpha)}^{q(\alpha)} f(x,\alpha)\,\mathrm{d}x$ *is differentiable on* $I$ *and*

$$F'(\alpha) =$$

$$= \int_{p(\alpha)}^{q(\alpha)} \frac{\partial f(x,\alpha)}{\partial \alpha} dx + q'(\alpha) f(q(\alpha),\alpha) - p'(\alpha) f(p(\alpha),\alpha). \quad (1.3.3)$$

* **Proof.** Take an arbitrary $\alpha_0 \in I$. We use the notations $p_0 = p(\alpha_0), q_0 = q(\alpha_0)$, splitting the integral (1.3.1) in three parts

$$F(\alpha) = \int_{p(\alpha)}^{q(\alpha)} f(x,\alpha) dx =$$

$$= \int_{p(\alpha)}^{p(\alpha_0)} f(x,\alpha) dx + \int_{p(\alpha_0)}^{q(\alpha_0)} f(x,\alpha) dx + \int_{q(\alpha_0)}^{q(\alpha)} f(x,\alpha) dx. \quad (1.3.4)$$

Let us now compute the ratio

$$\frac{F(\alpha) - F(\alpha_0)}{\alpha - \alpha_0} =$$

$$= \frac{1}{\alpha - \alpha_0} \int_{p(\alpha)}^{p_0} f(x,\alpha) dx + \int_{p_0}^{q_0} \frac{f(x,\alpha) - f(x,\alpha_0)}{\alpha - \alpha_0} dx + \quad (1.3.5)$$

$$+ \frac{1}{\alpha - \alpha_0} \int_{q_0}^{q(\alpha)} f(x,\alpha) dx.$$

We take by turns the integrals from the right member; we have

$$\frac{1}{\alpha - \alpha_0} \int_{p(\alpha)}^{p_0} f(x,\alpha) dx = \frac{p_0 - p(\alpha)}{\alpha - \alpha_0} f(\xi,\alpha), \ \xi \in (p(\alpha), p_0), \quad (1.3.6)$$

$$\frac{1}{\alpha - \alpha_0} \int_{q_0}^{q(\alpha)} f(x,\alpha) dx = \frac{q(\alpha) - q_0}{\alpha - \alpha_0} f(\eta,\alpha), \ \eta \in (q(\alpha), q_0), \quad (1.3.7)$$

by Lagrange's formula. We also have

$$\lim_{\alpha \to \alpha_0} \int_{p_0}^{q_0} \frac{f(x,\alpha) - f(x,\alpha_0)}{\alpha - \alpha_0} dx \overset{Th.1.15}{=}$$

$$= \int_{p_0}^{q_0} \lim_{\alpha \to \alpha_0} \frac{f(x,\alpha) - f(x,\alpha_0)}{\alpha - \alpha_0} dx = \int_{p_0}^{q_0} \frac{\partial f}{\partial \alpha}(x,\alpha_0) dx.$$

(1.3.8)

Further, we compute the limits

$$\lim_{\alpha \to \alpha_0} \frac{p_0 - p(\alpha)}{\alpha - \alpha_0} f(\xi,\alpha) = -\lim_{\alpha \to \alpha_0} \frac{p(\alpha) - p(\alpha_0)}{\alpha - \alpha_0} f(\xi,\alpha) =$$

$$= \underline{\underline{-p'(\alpha_0) f(p(\alpha_0),\alpha_0)}},$$

$$\lim_{\alpha \to \alpha_0} \frac{q(\alpha) - q_0}{\alpha - \alpha_0} f(\eta,\alpha) = \lim_{\alpha \to \alpha_0} \frac{q(\alpha) - q(\alpha_0)}{\alpha - \alpha_0} f(\eta,\alpha) =$$

$$= \underline{\underline{q'(\alpha_0) f(q(\alpha_0),\alpha_0)}}$$

(1.3.9)

The right member of the relation (1.3.5) makes sense if $\alpha$ goes to $\alpha_0$, whence it follows that the left member is also valid.

Consequently, we have

$$F(\alpha_0) = \lim_{\alpha \to \alpha_0} \frac{F(\alpha) - F(\alpha_0)}{\alpha - \alpha_0}.$$

(1.3.10)

By replacing the double underlined expressions from above in (1.3.5), after letting $\alpha \to \alpha_0$, we find

$$F'(\alpha_0) = \int_{p_0}^{q_0} \frac{\partial f}{\partial \alpha}(x,\alpha_0) dx + q'(\alpha_0) f(q(\alpha_0),\alpha_0) -$$

$$- p'(\alpha_0) f(p(\alpha_0),\alpha_0).$$

(1.3.11)

As $\alpha_0 \in I$ was arbitrary, we obtain (1.3.3). $\blacksquare$

*Example.* Compute the integral

$$I(\alpha, x) = \int_0^x \frac{dt}{\left(t^2 + \alpha^2\right)^2}. \qquad (1.3.12)$$

**Solution.**    Consider    the    auxiliary    integral

$$J(\alpha, x) = \int_0^x \frac{dt}{t^2 + \alpha^2}, \text{ which is easily computed}$$

$$J(\alpha, x) = \frac{1}{\alpha^2} \int_0^x \frac{dt}{1 + \left(\dfrac{t}{\alpha}\right)^2} = \frac{\alpha}{\alpha^2} \int_0^x \frac{d\left(\dfrac{t}{\alpha}\right)}{1 + \left(\dfrac{t}{\alpha}\right)^2} = \frac{1}{\alpha} \arctan \frac{x}{\alpha}. \quad (1.3.13)$$

$J(\alpha, x)$ satisfies the hypothesis of the theorem 1.16, therefore we can differentiate it with respect to $\alpha$, applying formula (1.3.3):

$$\frac{\partial J}{\partial \alpha} = \int_0^x \frac{\partial}{\partial \alpha}\left(\frac{1}{t^2 + \alpha^2}\right) dt = \int_0^x -\frac{2\alpha}{\left(t^2 + \alpha^2\right)^2} dt = -2\alpha I(\alpha, x). (1.3.14)$$

This can also be written in the form

$$I(\alpha, x) = -\frac{1}{2\alpha} \cdot \frac{\partial J}{\partial \alpha} = -\frac{1}{2\alpha} \cdot \frac{\partial}{\partial \alpha}\left[\frac{1}{\alpha} \arctan \frac{x}{\alpha}\right] =$$

$$= -\frac{1}{2\alpha}\left[-\frac{1}{\alpha^2} \arctan \frac{x}{\alpha} + \frac{1}{\alpha} \cdot \frac{1}{1 + \dfrac{x^2}{\alpha^2}} \cdot \left(-\frac{x}{\alpha^2}\right)\right]. \qquad (1.3.15)$$

Finally,

$$\boxed{I(\alpha, x) = \frac{1}{2\alpha^3} \arctan \frac{x}{\alpha} + \frac{x}{2\alpha^2\left(x^2 + \alpha^2\right)}.} \qquad (1.3.16)$$

# 1.4. NUMERICAL INTEGRATION

Many definite integrals cannot be analytically computed by using primitives. Moreover, in experiments, the integrand is often known only through data points. Therefore, it is natural to consider the numerical integration, i.e., to find appropriate methods in order to approximate the value of an integral of the type

$$J(f) = \int_a^b f(x)\mathrm{d}x. \tag{1.4.1}$$

## 1.4.1. THE TRAPEZIUM RULE

It is known that the integral (1.4.1) corresponds geometrically to the area comprised under the graph of the function $f(x)$ and the interval $[a,b]$. If $f(x)$ does not vary too much in the interval $[a,b]$, then this area can be approximated through the area of a trapezium ( figure 1.3).

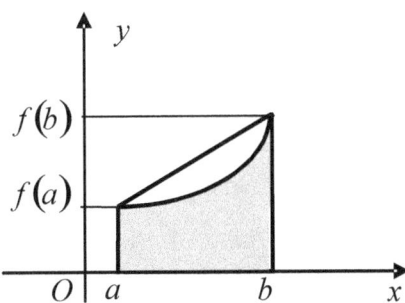

*Figure 1.3. The first trapezoidal formula (geometrical interpretation)*

Therefore, we have

$$\boxed{\int_a^b f(x)\mathrm{d}x \cong \frac{b-a}{2}\left[f(a)+f(b)\right]}.$$  (1.4.2)

Formula (1.4.2) is also known as the ***trapezium formula*** or ***rule***(in fact, ***the first trapezium formula***).

Another, less useful, alternative is to approximate the same area by the area of another trapezium, which is equivalent to the area of the rectangle from figure 1.4. We obtain ***the second trapezium formula,*** or ***the rectangle formula:***

$$\boxed{.\int_a^b f(x)\mathrm{d}x \cong (b-a)f\left(\frac{a+b}{2}\right)}.$$  (1.4.3)

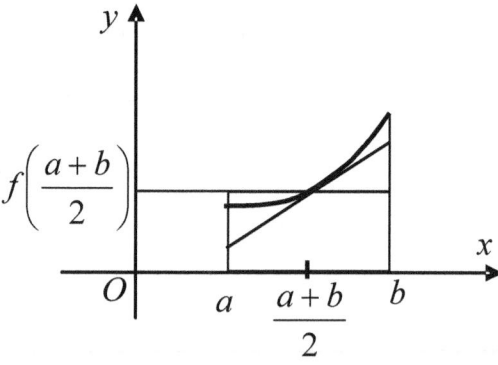

*Figure 1.4. The second trapezium rule (geometrical interpretation)*

The approximating formulae (1.4.2) and (1.4.3) are called ***(numeric) quadrature formulas***.

The difference $R_T(f)$ between the true and the approximate value of the integral is called ***the remainder*** of the quadrature formula (we used the index $T$ for "trapezia"). For example, for formula (1.4.2),

$$R_T(f) \equiv \int_a^b f(x)dx - \frac{b-a}{2}\left[f(a)+f(b)\right]. \qquad (1.4.4)$$

Actually, we can determine this remainder for each of the above-mentioned quadrature formulas. Let us take formula (1.4.2). Indeed, the following lemma holds true.

**Lemma 1.1.** *If* $f \in C^2\left(\left[a,b\right]\right)$, *then*

$$\left|\int_a^b f(x)dx - \frac{b-a}{2}\left[f(a)+f(b)\right]\right| \leq M\frac{|b-a|^3}{12},$$

$$M = \sup_{x\in[a,b]}\left|f''(x)\right|. \qquad (1.4.5)$$

\* **Proof.** Integrating (1.4.1) by parts, we get

$$\int_a^b f(x)dx =$$

$$= \int_a^b f(x)d(x-a) = f(b)(b-a) - \int_a^b f'(x)(x-a)dx. \qquad (1.4.6)$$

But

$$\int_a^b f'(x)(x-a)dx = \int_a^b f'(x)(x-a)d(x-b) =$$

$$= -\int_a^b (x-b)d\left[f'(x)(x-a)\right] = \qquad (1.4.7)$$

$$= -\int_a^b f''(x)(x-a)(x-b)dx - \int_a^b f'(x)(x-b)dx.$$

We integrate again, by parts, the last integral:

$$\int_a^b f'(x)(x-b)dx = f(a)(b-a) - \int_a^b f(x)dx. \qquad (1.4.8)$$

Replacing this in (1.4.7), we obtain:

$$\int_a^b f'(x)(x-a)\,dx = -\int_a^b f''(x)(x-a)(x-b)\,dx - $$

$$-f(a)(b-a)+\int_a^b f(x)\,dx. \qquad (1.4.9)$$

Finally, getting back to (1.4.6), it follows that

$$\int_a^b f(x)\,dx = \int_a^b f''(x)(x-a)(x-b)\,dx + $$

$$+(b-a)\big[f(a)+f(b)\big]-\int_a^b f(x)\,dx, \qquad (1.4.10)$$

which yields

$$\int_a^b f(x)\,dx - \frac{(b-a)}{2}\big[f(a)+f(b)\big] = $$

$$= \frac{1}{2}\int_a^b f''(x)(x-a)(x-b)\,dx. \qquad (1.4.11)$$

In the integral from the right member, we use Lagrange's formula and it results

$$\int_a^b f''(x)(x-a)(x-b)\,dx = -f''(c)\int_a^b (x-a)(x-b)\,dx = $$

$$= -f''(c)\frac{(b-a)^3}{6}. \qquad (1.4.12)$$

From this, we immediately get

$$R_T(f) = \frac{1}{2}\int_a^b f''(x)(x-a)(x-b)\,dx = -f''(c)\frac{(b-a)^3}{12}, \qquad (1.4.13)$$

whence the inequality (1.4.5) follows at once. $\blacksquare$

*Example.* Approximate the integral $\int_1^3 \dfrac{dx}{(x+1)^3}$ by using the first trapezoidal formula.

**Solution.** Comparing the data with formula (1.4.2), we note that in this case we have $a = 1$, $b = 3$, $f(x) = \dfrac{1}{(x+1)^3}$. It follows that

$$\int_1^3 \frac{dx}{(x+1)^3} \cong \frac{3-1}{2}\left(\frac{1}{2^3} + \frac{1}{4^3}\right) = \underline{0.140625}. \qquad (1.4.14)$$

This integral was considered on purpose, in order have the possibility to compare its approximate value with its value obtained by using primitives. The last one is easily obtained; indeed, we have

$$\int_1^3 \frac{dx}{(x+1)^3} \cong -\frac{1}{2(x+1)^2}\Big|_{x=1}^{x=3} = -\frac{1}{2}\left(\frac{1}{16} - \frac{1}{4}\right) = \underline{0.09375}. \quad (1.4.15)$$

Even without computing the remainder, we can notice the big difference between the true value and the approximate one.

## THE IMPROVED TRAPEZIUM RULE

However, we can improve the trapezium formula. Let us divide the interval $[a,b]$ into $n$ equal subintervals through $(n+1)$ equidistant points: $a = x_0, x_1, \ldots, x_n = b$. The length of each subinterval is $h = \dfrac{b-a}{n}$.

We have

$$\int_a^b f(x)\,dx = \sum_{i=1}^{n} \int_{x_{i-1}}^{x_i} f(x)\,dx = \sum_{i=1}^{n} J_i(f). \qquad (1.4.16)$$

Using the trapezoidal formula on each subinterval, we obtain

$$J_i(f) = \frac{x_i - x_{i-1}}{2}\left[f(x_i) + f(x_{i-1})\right] + R_{Ti} =$$
$$= \frac{h}{2}\left[f(x_i) + f(x_{i-1})\right] + R_{Ti}. \qquad (1.4.17)$$

We replace this in (1.4.16) and it results

$$J(f) = \frac{h}{2}\sum_{i=1}^{n}\left[f(x_i) + f(x_{i-1})\right] + \sum_{i=1}^{n} R_{Ti}, \qquad (1.4.18)$$

whence

$$\boxed{J(f) \cong \frac{h}{2}\sum_{i=1}^{n}\left[f(x_i) + f(x_{i-1})\right]}. \qquad (1.4.19)$$

This formula is also known as *the improved trapezoidal formula*.

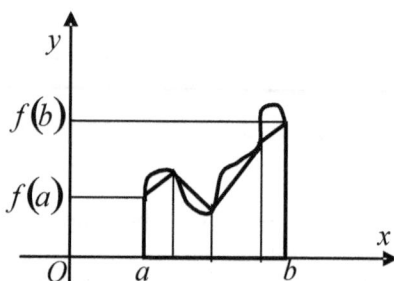

*Figure 1.5. The improved trapezium formula. Geometrical interpretation*

Taking into account that $h = \dfrac{b-a}{n}$ and $x_i = a + ih$, $i = \overline{0, n-1}$, this formula can be also written as:

$$J(f) = \frac{b-a}{2n}\left[f(a) + 2f(a+h) + 2f(a+2h) + \right.$$
$$\left. + \ldots + 2f[a+(n-1)h] + f(b)\right]. \tag{1.4.20}$$

From lemma 1.1 it follows that, if $f''(x)$ is bounded on $[a,b]$, i.e. $|f''(x)| < M$, $x \in [a,b]$, then each remainder $R_{Ti}$ can be majorized on each interval $[x_{i-1}, x_i]$ by

$$|R_{Ti}(f)| \le M \frac{(x_i - x_{i-1})^3}{12}, \tag{1.4.21}$$

or

$$|R_{Ti}(f)| \le M \frac{h^3}{12}. \tag{1.4.22}$$

Summing up these inequalities for $i = \overline{1,n}$ and taking into account the relations (1.4.18) and (1.4.21), we deduce

$$\left|\sum_{i=1}^{n} R_{Ti}\right| \le \sum_{i=1}^{n} |R_{Ti}| \le n \cdot M \frac{h^3}{12} = n \cdot \frac{(b-a)^3}{n^3} \cdot \frac{M}{12}, \tag{1.4.23}$$

or

$$|R_T(f)| < M \frac{(b-a)^3}{12n^2}, \quad M = \sup_{x \in [a,b]} |f''(x)|. \tag{1.4.24}$$

By using the remainder formula, one can compute how many intervals are necessary in order to approximate a given integral with a given precision. For example, if we want to compute the integral with a precision of 3 decimal places, then the following condition must be fulfilled

$$M \frac{(b-a)^3}{12n^2} < 10^{-3} \Rightarrow n > \sqrt{M \frac{(b-a)^3}{12 \cdot 10^{-3}}} \equiv \sigma. \qquad (1.4.25)$$

It suffices to choose $n = [\sigma] + 1$ in order to assure the precision of 3 decimal places ($[\ ]$ is the notation for the integer part).

*Example.* Let us reconsider the integral $\int\limits_1^3 \frac{dx}{(x+1)^3}$. We wish to compute it with a precision of $\varepsilon = 0.02$.

**Solution.** We establish the number of intervals necessary to the required precision. By computing

$$f' = -3(x+1)^{-4}, f'' = 12(x+1)^{-5},$$

we get $M = \frac{12}{2^5} = \frac{3}{8}$. Therefore $n > \sqrt{\frac{3}{8} \cdot \frac{2^3}{12 \cdot 0.02}} \cong 3.53$. It suffices to take $n = 4$. Thus, the step of the network is $h = \frac{b-a}{n} = \frac{3-1}{4} = 0.5$, hence $\delta = \{1, 1.5, 2, 2.5, 3\}$. The values of the function at the points of $\delta$ are given in the following table:

| $x$ | 1 | 1.5 | 2 | 2.5 | 3 |
|---|---|---|---|---|---|
| $f(x)$ | 0.1250 | 0.0640 | 0.0370 | 0.0233 | 0.0156 |

We use the improved trapezoidal formula:

$$\int\limits_1^3 \frac{dx}{(x+1)^3} \cong \frac{0.5}{2} \sum_{j=1}^4 \left[ f(x_{j-1}) + f(x_j) \right] =$$

$$= 0.25 \left[ f(1) + 2f(1.5) + 2f(2) + 2f(2.5) + f(3) \right] = \qquad (1.4.26)$$

$$= 0.25 \left[ 0.125 + 2 \cdot (0.064 + 0.0370 + 0.0233) + 0.0156 \right],$$

which finally gives

$$\int_{1}^{3} \frac{dx}{(x+1)^3} \cong \underline{0.09732.}$$

The true value is, as shown before, $\underline{0.09375.}$

## 1.4.2. SIMPSON'S RULE

Let us notice that the first trapezium rule was based on approximating the integrand $f(x)$ by the segment which joins the points of coordinates $(a, f(a)), (b, f(b))$. We can obtain a more refined approximation if we approximate the integrand by a parabola of equation

$$y = Ax^2 + Bx + C, \qquad (1.4.27)$$

passing through three points, let them be:

$$(a, f(a)), (b, f(b)), \left(\frac{a+b}{2}, f\left(\frac{a+b}{2}\right)\right), \qquad (1.4.28)$$

$\frac{a+b}{2}$ being the midpoint of the segment $[a, b]$. This yields the following equalities:

$$A a^2 + B a + C = f(a),$$
$$A\left(\frac{a+b}{2}\right)^2 + B\frac{a+b}{2} + C = f\left(\frac{a+b}{2}\right), \qquad (1.4.29)$$
$$A b^2 + B b + C = f(b).$$

The integral is then approximated by

65

$$J(f) \equiv \int_a^b f(x)\,dx \cong \int_a^b \left(Ax^2 + Bx + C\right)dx. \qquad (1.4.30)$$

From (1.4.29), we obtain immediately

$$A = 2\frac{f(a) + f(b) - 2f\left(\dfrac{a+b}{2}\right)}{(b-a)^2}. \qquad (1.4.31)$$

We compute the approximation:

$$
\begin{aligned}
\int_a^b \left(Ax^2 + Bx + C\right)dx &= A\frac{b^3 - a^3}{3} + B\frac{b^2 - a^2}{2} + C(b-a) = \\
&= \frac{(b-a)}{6}\left[2A\left(a^2 + ab + b^2\right) + 3B(a+b) + 6C\right] = \qquad (1.4.32) \\
&= \frac{(b-a)}{3}\left[f(a) + f(b) + Aab + B\frac{a+b}{2} + C\right].
\end{aligned}
$$

But from the relations (1.4.29) we obtain

$$B(a+b) + 2C = 2f\left(\frac{a+b}{2}\right) - \frac{A}{2}\left(a^2 + ab + b^2\right), \qquad (1.4.33)$$

and, after some elementary calculation, we get

$$
\begin{aligned}
\int_a^b \left(Ax^2 + Bx + C\right)dx &= \\
&= \frac{(b-a)}{6}\left[f(a) + f(b) + 4f\left(\frac{a+b}{2}\right)\right].
\end{aligned} \qquad (1.4.34)
$$

Finally,

$$J(f) \equiv \int_a^b f(x)\,dx \cong \frac{(b-a)}{6}\left[f(a) + f(b) + 4f\left(\frac{a+b}{2}\right)\right]. \quad (1.4.35)$$

This is **Simpson's quadrature formula**.

66

The remainder is estimated analogously to that for trapezia. We obtain the inequality (similar to that from lemma 1.1):

$$\left| \int_a^b f(x)dx - \frac{(b-a)}{6}\left[ f(a)+f(b)+4f\left(\frac{a+b}{2}\right) \right] \right| \le$$

$$\le K\frac{(b-a)^5}{2880}, \quad K = \sup_{x\in[a,b]} f^{(4)}(x).$$

(1.4.36)

Hence, the remainder $R_S(f)$ of Simpson's formula satisfies the inequality

$$R_S(f) \le K\frac{(b-a)^5}{2880}, \qquad K = \sup_{x\in[a,b]} f^{(4)}(x).$$

(1.4.37)

*Example.* Approximate the above integral $\int_1^3 \frac{dx}{(x+1)^3}$ by using Simpson's formula.

**Solution.** In (1.4.35) we take $a = 1$, $b = 3$, $f(x) = \frac{1}{(x+1)^3}$,

$$\frac{a+b}{2} = 2.$$

We approximate the integral using Simpson's formula:

$$\int_1^3 \frac{dx}{(x+1)^3} \cong \frac{3-1}{6}\left( \frac{1}{2^3} + 4\cdot\frac{1}{3^3} + \frac{1}{4^3} \right) = \underline{0.09626}.$$

(1.4.38)

As previously shown, the true value is $\underline{0.09375}$.

# THE IMPROVED SIMPSON'S FORMULA

Like the trapezium formula, Simpson's formula can be improved by dividing the interval $[a,b]$ in $2n$ equal subintervals, through the equidistant partition points

$$a = x_0 < x_1 \ldots < x_{2i-2} < x_{2i-1} < x_{2i} < \ldots x_{2n} = b.$$

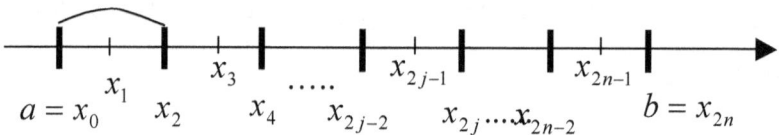

*Figure 1.6. The network for the improved Simpson's formula*

The step of this network is thus $h = \dfrac{b-a}{2n}$, therefore the arc from the above figure is subtended by a segment of length $2h$.

We can write

$$J(f) = \int_a^b f(x)dx = \sum_{i=1}^n \int_{x_{2i-2}}^{x_{2i}} f(x)dx. \tag{1.4.39}$$

If we use Simpson's formula on each subinterval $[x_{2i-2}, x_{2i}]$, for the points $x_{2i-2}$, $x_{2i-1}$, $x_{2i}$, we obtain

$$
\int_{x_{2i-2}}^{x_{2i}} f(x)dx =
$$

$$
= \frac{x_{2i} - x_{2i-2}}{6}\left[f(x_{2i-2}) + 4f(x_{2i-1}) + f(x_{2i})\right] + R_{3i}. \tag{1.4.40}
$$

But $\dfrac{x_{2i} - x_{2i-2}}{6} = \dfrac{2h}{6} = \dfrac{h}{3}$, hence

$$J(f) = \frac{h}{3}\sum_{i=1}^{n}\left[f\left(x_{2i-2}\right)+4f\left(x_{2i-1}\right)+f\left(x_{2i}\right)\right]+\sum_{i=1}^{n}R_{3i}. \quad (1.4.41)$$

Consequently, the integral can be approximated using the *improved Simpson's formula*

$$\boxed{J(f) \cong \frac{h}{3}\sum_{i=1}^{n}\left[f\left(x_{2i-2}\right)+4f\left(x_{2i-1}\right)+f\left(x_{2i}\right)\right],} \quad (1.4.42)$$

and the remainder of this formula is

$$R_{S}(f) = \sum_{i=1}^{n}R_{3i}. \quad (1.4.43)$$

If $K = \sup\limits_{x\in[a,b]} f^{(4)}(x)$, then each $R_{3i}$ satisfies the inequality (1.4.37), on the corresponding interval $\left[x_{2i-2}, x_{2i}\right]$. We infer that

$$\left|R_{S}(f)\right| \leq \sum_{i=1}^{n}\left|R_{3i}\right| \leq \sum_{i=1}^{n}K\frac{\left|x_{2i}-x_{2i-1}\right|^{5}}{2880} =$$
$$= K\sum_{i=1}^{n}\frac{h^{5}}{2880} = Kn\cdot\frac{(b-a)^{5}}{2880n^{5}}. \quad (1.4.44)$$

It follows that the remainder of Simpson's formula satisfies the inequality

$$\left|R_{S}(f)\right| < K\frac{(b-a)^{5}}{2880n^{4}}. \quad (1.4.45)$$

*Remark.* By comparing the inequalities (1.4.24) and (1.4.45), we notice that Simpson's formula provides a better approximant than the trapezium formula, for a smaller number of subintervals.

As in the case of the improved trapezium rule, we can establish the number of intervals in order to obtain the value of the integral with a given precision. For example, if we want to compute the integral with a precision of 3 decimal places, then the following condition must be fulfilled

$$K\frac{(b-a)^5}{2880n^4} < 10^{-3} \quad \Rightarrow \quad n \geq \sqrt[4]{K\frac{(b-a)^5}{2880 \cdot 10^{-3}}} \equiv \sigma. \quad (1.4.46)$$

It is suffices to choose $n = [\sigma] + 1$ in order to assure the precision of 3 decimals places ([ ] is the notation for the integer part).

*Example.* Consider again the integral $\int_1^3 \frac{dx}{(x+1)^3}$ and let us compute it with a precision of $\varepsilon = 0.02$.

**Solution.** Firstly we establish the number of intervals necessary to the required precision. By computing

$$f'''(x) = -60(x+1)^{-6}, \quad f^{(4)}(x) = 360(x+1)^{-7},$$

we get $M = \dfrac{360}{2^7} = \dfrac{45}{16} = 2.1825$.

Therefore $n > \sqrt[4]{\dfrac{45}{2^4} \cdot \dfrac{2^5}{2880 \cdot 0.02}} \cong 1.12$. It suffices to take $n = 1$. The network has the step $h = \dfrac{b-a}{2n} = \dfrac{3-1}{2} = 1$, therefore $\delta = \{1, 2, 3\}$. We have previously computed the approximate value of the integral and we obtained

$$\int_1^3 \frac{dx}{(x+1)^3} \cong \underline{0.09626}.$$

Now, let us use the same network as for the improved trapezium formula, i.e., $\delta = \{1, 1.5, 2, 2.5, 3\}$; the number of points was established there for the precision of $\varepsilon = 0.02$.

The following table contains the values of the function at the points of $\delta$:

| $x$ | 1 | 1.5 | 2 | 2.5 | 3 |
|-----|-----|------|--------|--------|--------|
| $f(x)$ | 0.125 | 0.064 | 0.0370 | 0.0233 | 0.0156 |

We use the improved Simpson's formula; here, $n = 2$:

$$\int_1^3 \frac{dx}{(x+1)^3} \cong \frac{0.5}{3} \sum_{j=1}^{2} \left[ f(x_{2i-2}) + 4f(x_{2i-1}) + f(x_{2i}) \right] =$$

$$= \frac{0.5}{3} \left[ f(1) + 4f(1.5) + 2f(2) + 4f(2.5) + f(3) \right] =$$

$$= \frac{0.5}{3} \left[ 0.125 + 4(0.064 + 0.0233) + 2 \cdot 0.0370 + 0.0156 \right],$$

which finally yields

$$\int_1^3 \frac{dx}{(x+1)^3} \cong \underline{0.0939666}.$$

The true value of the integral was computed above: 0.09375. Its approximating value by using the improved trapezium rule on $\delta$ was 0.09732.

The remainder estimation is, according to formula (1.4.45)

71

$$\left| R_3 \right| < K \frac{(b-a)^5}{2880n^4} = \frac{45}{2^4} \cdot \frac{2^5}{2880 \cdot 2^4} \cong 0.00195 < 0.002.$$

At the fourth decimal only one can notice a difference between the approximate and the true value!

This example emphasizes that

**Simpson's method is more accurate than the trapezium method.**

## EXERCISES AND PROBLEMS

1. Compute the following integrals and primitives.

**A.** Integrals with trigonometric integrand:

a) $I = \int \dfrac{dx}{\sin^2 x \cos^2 x}$, $x \in \left( 0, \dfrac{\pi}{2} \right)$  A: $I = \tan x - \cot x + C$

b) $I = \int \sin^5 x \cos x \, dx$, $x \in \Re$   A: $I = \dfrac{1}{6} \sin^6 x + C$

c) $I = \int \dfrac{\sin x}{1 + \sin^2 x} dx$, $x \in \left[ 0, \dfrac{\pi}{2} \right]$  A: $I = \dfrac{1}{2\sqrt{2}} \ln \left| \dfrac{\cos x - \sqrt{2}}{\cos x + \sqrt{2}} \right| + C$

d) $I = \int \dfrac{\cos^3 x}{\sin x} dx$, $x \in \left( 0, \dfrac{\pi}{2} \right]$  A: $I = \ln \left| \sin x \right| - \dfrac{\sin^2 x}{2} + C$

e) $I = \int \dfrac{\sin x}{(1 - \cos x)^3} dx$, $x \in \left( 0, \dfrac{\pi}{2} \right]$ A: $I = -\dfrac{1}{2(\cos x - 1)^2} + C$

f) $I = \int \dfrac{\cos x}{\sin^2 x - 6 \sin x + 5} dx$, $x \in \left[ 0, \dfrac{\pi}{2} \right)$  A: $I = \dfrac{1}{4} \ln \left| \dfrac{\sin x - 5}{\sin x - 1} \right| + C$

g) $I = \int \dfrac{dx}{\sin x + 2\cos x + 3}$

$$\text{A: } I = \arctan\left(\dfrac{\tan\dfrac{x}{2} + 1}{2}\right) + C$$

*Hint:* $\quad t = \tan\dfrac{x}{2}, \ \sin x = \dfrac{2t}{1+t^2},$

$$\cos x = \dfrac{1-t^2}{1+t^2}$$

h) $I = \int \dfrac{dx}{\sin x}, \ x \in (0, \pi)$ $\qquad$ A: $I = \ln\left|\tan\dfrac{x}{2}\right| + C$

i) $I = \int \dfrac{dx}{5 - 3\cos x}, \ x \in \Re$ $\qquad$ A: $I = \dfrac{1}{4}\arctan\left(2\tan\dfrac{x}{2}\right) + C$

j) $I = \int \dfrac{dx}{(2 - \sin x)(3 - \sin x)}$

$$\text{A: } I = \dfrac{2}{\sqrt{3}}\arctan\dfrac{2\tan\dfrac{x}{2} - 1}{\sqrt{3}} - \dfrac{1}{\sqrt{2}}\arctan\dfrac{3\tan\dfrac{x}{2} - 1}{2\sqrt{2}} + C$$

k) $I = \int \dfrac{dx}{\sin x + \cos x}$ $\qquad$ A: $I = -\dfrac{1}{\sqrt{2}}\ln\left|\dfrac{\tan\dfrac{x}{2} - 1 - \sqrt{2}}{\tan\dfrac{x}{2} - 1 + \sqrt{2}}\right| + C$

l) $I = \int \dfrac{1 - \sin x + \cos x}{1 + \sin x - \cos x}dx$ $\qquad$ A: $I = \ln\dfrac{\left|\tan\dfrac{x}{2}\right|}{\left(\tan\dfrac{x}{2} + 1\right)^2} + C$

m) $\displaystyle\int \frac{\sin x\, dx}{e^x + \sin x + \cos x}$   A: $I = \dfrac{1}{2}\Big[x - \ln\big|e^x + \sin x + \cos x\big|\Big] + C$

$$Hint:\ J = \frac{e^x + \cos x}{e^x + \sin x + \cos x}$$

n) $I = \displaystyle\int \frac{dx}{(\sin x + \cos x)^2}$   A: $I = -\dfrac{1}{\tan x + 1}$

$$Hint:\ \tan x = t$$

o) $I = \displaystyle\int \frac{2\tan x + 3}{\sin^2 x + 2\cos^2 x}\, dx$

A: $I = \ln\big(\tan^2 x + 2\big) + \dfrac{3}{\sqrt{2}}\arctan\dfrac{\tan x}{\sqrt{2}} + C$

p) $I = \displaystyle\int \frac{dx}{3\sin^2 x - 8\sin x \cos x + 5\cos^2 x}$

A: $I = \dfrac{1}{2}\ln\left|\dfrac{3\sin x - 5\cos x}{\sin x - \cos x}\right| + C$

r) $I = \displaystyle\int \frac{3\sin x + 2\cos x}{2\sin x + 3\cos x}\, dx$

A: $I = -\dfrac{5}{13}\ln\big|2\sin x + 3\cos x\big| + \dfrac{12}{13}x + C$

s) $I = \displaystyle\int \frac{dx}{1 + 3\cos^2 x}$   A: $I = \dfrac{1}{2}\arctan\dfrac{\tan x}{2} + C$

t) $I = \displaystyle\int \frac{dx}{3\sin^2 x + 5\cos^2 x}$   A: $I = \dfrac{1}{\sqrt{15}}\arctan\left(\sqrt{\dfrac{3}{5}}\tan x\right) + C$

u) $I = \displaystyle\int \frac{dx}{\sin^2 x + 3\sin x \cos x - \cos^2 x}$

A: $I = \dfrac{1}{\sqrt{13}}\ln\left|\dfrac{2\tan x + 3 - \sqrt{13}}{2\tan x + 3 + \sqrt{13}}\right| + C$

**B.** Integrals with irrational integrand:

a) $\int \dfrac{dx}{\sqrt{2x-1} - \sqrt[3]{2x-1}}, \; x \in (1, \infty)$

$$A: I = \sqrt{2x-1} + \dfrac{3}{2}\sqrt[3]{2x-1} + 3\sqrt[6]{2x-1} +$$
$$+ 3\ln\left(\sqrt[6]{2x-1} - 1\right) + C$$

**C.** Integrals with irrational integrand, using Euler's substitutions:

a) $I = \int \dfrac{dx}{(x+1)\sqrt{x^2+x+1}}, \; x \in (-1, \infty)$

$$A: I = \ln\left| \dfrac{\sqrt{x^2+x+1} - x - 2}{\sqrt{x^2+x+1} - x} \right| + C$$

Hint: $\sqrt{x^2+x+1} = x + t$

b) $I = \int \dfrac{x+2}{\sqrt{x(x-3)}}dx,$

$\quad x \in (3, \infty)$

$$A: I = \dfrac{7}{2}\ln\left| \dfrac{1 + \sqrt{\dfrac{x-3}{x}}}{1 - \sqrt{\dfrac{x-3}{x}}} \right| + x\sqrt{\dfrac{x-3}{x}} + C$$

Hint: $\sqrt{x(x-3)} = tx$

c) $I = \int \dfrac{dx}{x^2\sqrt{x^2-x+1}},$

$\quad x \in (0, \infty)$

$$A: I = -\dfrac{\sqrt{x^2-x+1}-1}{2x} +$$
$$+ \dfrac{1}{2}\ln\left| \dfrac{2\sqrt{x^2-x+1}-2+x}{x} \right| -$$
$$- \dfrac{3}{2}\dfrac{x}{2\sqrt{x^2-x+1}-2+x} + C$$

Hint: $\sqrt{x^2-x+1} = tx+1$

d) $I = \int \dfrac{dx}{1 + \sqrt{-x^2 + 3x - 2}}$  $\quad A: I = -2\arctan\sqrt{\dfrac{2-x}{x-1}} +$

$$+ \dfrac{4}{\sqrt{3}}\arctan\dfrac{1}{\sqrt{3}}\left(2\sqrt{\dfrac{2-x}{x-1}} + 1\right) + C$$

Hint: $\sqrt{-(x-1)(x-2)} = t\,(x-1)$

e) $I = \int \dfrac{dx}{(x+1)\sqrt{x^2 + 2x - 3}}, \quad x \in (1, \infty)$

$$A: I = -\arctan\sqrt{\dfrac{x+3}{x-1}} + C$$

Hint: $\sqrt{(x-1)(x+3)} = t\,(x-1)$

2. Starting from the definintion, establish the nature of the following improper integrals on unbounded intervals and compute them, in case of convergence:

a) $I = \displaystyle\int_0^\infty \sin\alpha x\,dx$  $\qquad$ A: $I$ is divergent

b) $I = \displaystyle\int_0^\infty \dfrac{x\,dx}{x^2 + 1}$  $\qquad$ A: $I$ is divergent

c) $I = \displaystyle\int_0^\infty \dfrac{dx}{x^2 + 1}$  $\qquad$ A: $I$ is convergent and $I = \dfrac{\pi}{2}$

d) $I = \displaystyle\int_0^\infty e^{-ax}dx, \; a > 0$  $\qquad$ A: $I$ is convergent and $I = \dfrac{1}{a}$

e) $I = \displaystyle\int_0^\infty \dfrac{x^2}{\left(x^3 + 1\right)^2}\,dx$  $\qquad$ A: $I$ is convergent and $I = \dfrac{1}{3}$

f) $I = \displaystyle\int_{-\infty}^1 \dfrac{2x+1}{x^2 + 1}\,dx$  $\qquad$ A: $I$ is divergent

3. Using the convergence criteria, establish the nature of the following improper integrals on unbounded intervals and find their values, in case of convergence:

a) $I = \int\limits_0^\infty \dfrac{dx}{\sqrt{5x+1}}$  A: $I$ is divergent

b) $I = \int\limits_2^\infty \dfrac{dx}{x^2 + x}$  A: $I$ is convergent and $I = -\ln\dfrac{2}{3}$

c) $I = \int\limits_1^\infty \dfrac{dx}{x(x^2+1)}$  A: $I$ is convergent and $I = \dfrac{1}{2}\ln 2$

d) $I = \int\limits_1^\infty \dfrac{dx}{x\sqrt{x^2+1}}$  A: $I$ is convergent and $I = \ln(1+\sqrt{2})$

e) $I = \int\limits_1^\infty \dfrac{dx}{\sqrt{x}(x+1)}$  A: $I$ is convergent and $I = \dfrac{\pi}{2}$

f) $I = \int\limits_1^\infty \dfrac{x\,dx}{(x^2+1)^2}$  A: $I$ is convergent and $I = \dfrac{1}{4}$

g) $I = \int\limits_{-\infty}^\infty \dfrac{x}{1+x^4}\,dx$  A: $I$ is convergent and $I = 0$

h) $I = \int\limits_0^\infty \dfrac{dx}{(x+a)(x+b)}$, $a > 0, b > 0$  A: $I$ is convergent and $I = \dfrac{1}{a-b}\ln\dfrac{a}{b}$

i) $I = \int\limits_0^\infty \dfrac{x^2 dx}{(x^2+a^2)(x^2+b^2)}$, $a \neq b$  A: $I$ is convergent and $I = \dfrac{\pi}{2}(b+a)$

j) $I = \int\limits_0^\infty \dfrac{x\,dx}{(x+a)(x^2+b^2)}$, $a > 0, b > 0$

A: $I$ is convergent and

$$I = \dfrac{1}{a^2+b^2}\left(\dfrac{b\pi}{2} - a\ln\dfrac{b}{a}\right)$$

k) $I = \int\limits_0^\infty e^{-x} \sin x \, dx$

A: $I$ is convergent and $I = \dfrac{1}{2}$

l) $I = \int\limits_0^\infty e^{-x} \cos kx \, dx$

A: $I$ is convergent and $I = \dfrac{1}{k^2 + 1}$

m) $I = \int\limits_0^\infty \dfrac{\arctan x}{1 + x^2} \, dx$

A: $I$ is convergent and $I = \dfrac{\pi^2}{8}$

n) $I = \int\limits_2^\infty \dfrac{1}{x \ln^2 x} \, dx$

A: $I$ is convergent and $I = \dfrac{1}{\ln 2}$

4. Study the absolute convergence of the following improper integrals on unbounded intervals:

a) $I = \int\limits_0^\infty \dfrac{\sin x}{a^2 + x^2} \, dx$    A: $I$ is absolutely convergent

b) $I = \int\limits_1^\infty \dfrac{\arctan x}{x^n} \, dx,$   $n > 1$   A: $I$ is absolutely convergent

5. Establish the nature of the following improper integrals with unbounded integrand and find their values, in case of convergence:

a) $I = \int\limits_0^1 \dfrac{dx}{\sqrt{x}}$    A: $I$ is convergent and $I = 2$

b) $I = \int\limits_0^1 \dfrac{dx}{(x-1)^2}$    A: $I$ is divergent

c) $I = \int\limits_0^{\frac{\pi}{2}} \dfrac{dx}{\sin x}$    A: $I$ is divergent

d) $I = \int\limits_0^{\frac{\pi}{2}} \cot x \, dx$    A: $I$ is divergent

e) $I = \int_{-1}^{1} \dfrac{dx}{x^2 - 1}$    A: $I$ is divergent

*Hint*: The function is unbounded

at $x = 1, x = -1$

f) $I = \int_{2}^{3} \dfrac{dx}{\sqrt[5]{x - 2}}$    A: $I$ is convergent and $I = \dfrac{5}{4}$

g) $I = \int_{0}^{a} \dfrac{x\,dx}{\sqrt{a^2 - x^2}}, \ a > 0$    A: $I$ is convergent and $I = a$

h) $I = \int_{0}^{a} \dfrac{dx}{\sqrt{a^2 - x^2}}, \ a > 0$  A: $I$ is convergent and $I = \dfrac{\pi}{2}$

6. Compute the following improper integrals using the functions $\Gamma$ and $\beta$ :

a) $\Gamma\left(\dfrac{9}{2}\right)$    A: $\Gamma\left(\dfrac{9}{2}\right) = \dfrac{105\sqrt{\pi}}{16}$

b) $\Gamma\left(\dfrac{2n+1}{2}\right)$    A: $\Gamma\left(\dfrac{2n+1}{2}\right) = \dfrac{(2n-1)!}{2^{2n-1}(n-1)!}\sqrt{\pi}$

c) $I = \int_{0}^{\infty} x^p e^{-x^q}\,dx, \ p > -1, q > 0$    A: $I = \dfrac{1}{q}\Gamma\left(\dfrac{p+1}{q}\right)$

d) $I = \int_{0}^{1} x^{-\frac{2}{3}}(1-x)^{-\frac{1}{3}}\,dx$    A: $I = \beta\left(\dfrac{1}{3}, \dfrac{2}{3}\right) = \dfrac{2\sqrt{3}\pi}{3}$

e) $I = \int_{0}^{\infty} x^2 (1+x)^{-4}\,dx$    A: $I = \beta(3,1) = \dfrac{1}{3}$

f) $I = \int_{0}^{\infty} \dfrac{dx}{1 + x^6}$    A: $I = \beta\left(\dfrac{1}{6}, \dfrac{5}{6}\right) = 2\pi$

*Hint* $: x^6 = t$

g) $I = \int_0^a x^2 \sqrt{a^2 - x^2} \, dx, \; a > 0$  $\qquad$ A: $I = \beta\left(\dfrac{3}{2}, \dfrac{3}{2}\right) = \dfrac{\pi}{8}$

$$\text{Hint: } x^2 = a^2 t$$

h) $I = \int_0^{\frac{\pi}{2}} \sin^3 x \cos^5 x \, dx,$  $\qquad$ A: $I = \dfrac{1}{24}$

i) $I = \int_0^1 x^{p-1} \left(1 - x^m\right)^{q-1} dx, \; p > 0, q > 0, m > 0$

$$\text{A: } I = \dfrac{1}{m}\beta\left(\dfrac{p}{m}, q\right)$$

7. Compute the following integrals, using the differentiation with respect to the parameter:

a) $I(\alpha) = \int_0^{\frac{\pi}{2}} \dfrac{\arctan(\alpha \tan x)}{\tan x} dx,$  $\qquad$ A: $I(\alpha) = \dfrac{\pi}{2}\ln|1 + \alpha|$

$\alpha > 0, \alpha \neq 1$

b) $I(\alpha) = \int_0^{\alpha} \dfrac{\ln(1 + \alpha x)}{1 + x^2} dx$  $\qquad$ A: $I(\alpha) = \dfrac{1}{2}\arctan\alpha \cdot \ln\left(1 + \alpha^2\right)$

c) $I(\alpha) = \int_0^{\frac{\pi}{2}} \ln\left(\alpha^2 - \sin^2 x\right) dx, \; a > 1$  $\qquad$ A: $I(\alpha) = \pi \ln\left|\dfrac{\alpha + \sqrt{\alpha^2 - 1}}{2}\right|$

8. Given: $M = \sup\limits_{x \in [a,b]} |f''(x)|$ and $K = \sup\limits_{x \in [a,b]} |f^{(4)}(x)|$,

approximate the following integrals with the precision $\varepsilon$, using, by comparison, the improved trapezium rule and Simpson's rule:

a) $I = \int_0^1 \dfrac{x}{1 + x^2} dx,$ $\quad$ $\begin{aligned} M &= 8, \\ K &= 192, \\ \varepsilon &= 0.005. \end{aligned}$ $\qquad$ b) $I = \int_0^{\frac{\pi}{2}} x \sin x \, dx,$ $\quad$ $\begin{aligned} M &= \pi/2 + 2, \\ K &= \pi/2 + 4, \\ \varepsilon &= 0.001. \end{aligned}$

# Chapter 2

# THE CURVILINEAR INTEGRAL

## 2.1. THE ARC LENGTH

We measure a segment by comparing it with another segment, taken as unit.

### *How do we proceed with an arc?*

Naturally, it can be straightened up; thus, it becomes a segment. This procedure is called ***rectification***.

### NOTIONS OF ELEMENTARY GEOMETRY

***The arch length*** is defined as the limit of the perimeters of the polygons inscribed in the arc, when the number of sides tends to infinity.

Consider the plane curve $\overset{\frown}{AB} \subset \Re^2$ of equations:

$$\overset{\frown}{AB} : \begin{cases} x = x(t) \\ y = y(t) \end{cases}, \quad t \in [a,b]. \tag{2.1.1}$$

We assume that $\overset{\frown}{AB}$ is smooth, i.e., $x, y \in C^1\left([a,b]\right)$.

Let us divide $[a,b]$ in $n$ subsegments, using the partition $\Delta = \{a = t_0, t_1, \ldots, t_n = b\}$. To each $t_k$ we associate the

81

corresponding point $M_k = M\left(x_k, y_k\right)$ belonging to the arc $\overset{\frown}{AB}$;

here, we used the notation $x_k = x\left(t_k\right)$, $y_k = y\left(t_k\right)$, $k = \overline{0, n}$.

We have $A\left(x\left(a\right), y\left(a\right)\right), B\left(x\left(b\right), y\left(b\right)\right)$ ( figure 2.1).
$$\underset{x_0}{\underbrace{x(a)}}\quad \underset{y_0}{\underbrace{y(a)}}\quad \underset{x_n}{\underbrace{x(b)}}\quad \underset{y_n}{\underbrace{y(b)}}$$

Let us join the points $M_k$, thus setting up the polygonal line $L_n \equiv AM_1M_2...M_{n-1}B$. Its length, denoted by $l_n$, tends with $n \to \infty$ to the length $l$ of the arc $\overset{\frown}{AB}$.

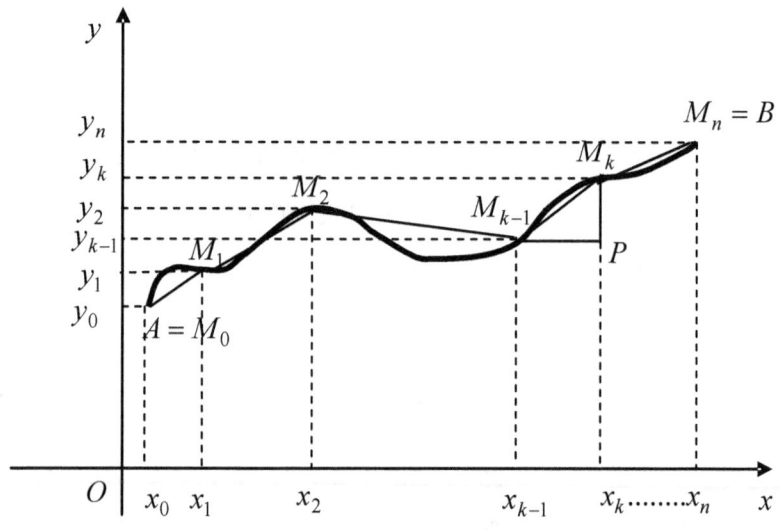

*Figure 2.1. The polygonal line inscribed in the arc*

We now compute $l_n$:

$$l_n = \sum_{k=1}^{n}\left|\overline{M_{k-1}M_k}\right|. \tag{2.1.2}$$

But

$$\left|\overline{M_{k-1}M_k}\right| = \sqrt{\left|\overline{M_kP}\right|^2 + \left|\overline{M_{k-1}P}\right|^2}, \tag{2.1.3}$$

and

$$\left|\overline{M_k P}\right| = \left|y_k - y_{k-1}\right| = \left|y\left(t_k\right) - y\left(t_{k-1}\right)\right| = \left|y'\left(\beta_k\right)\right|\left|t_k - t_{k-1}\right|,$$
(2.1.4)

$$\beta_k \in \left(t_{k-1}, t_k\right),$$

by Lagrange's formula (see [4,10]).

Analogously, we have

$$\left|\overline{M_{k-1} P}\right| = \left|x_k - x_{k-1}\right| = \left|x\left(t_k\right) - x\left(t_{k-1}\right)\right| = \left|x'\left(\alpha_k\right)\right|\left|t_k - t_{k-1}\right|,$$
(2.1.5)

$$\alpha_k \in \left(t_{k-1}, t_k\right)$$

again due to Lagrange's formula.

Therefore

$$\left|\overline{M_{k-1} M_k}\right| = \left|t_k - t_{k-1}\right| \cdot \sqrt{\left[x'\left(\alpha_k\right)\right]^2 + \left[y'\left(\beta_k\right)\right]^2}.$$
(2.1.6)

This gives

$$l_n = \sum_{k=1}^{n} \sqrt{\left[x'\left(\alpha_k\right)\right]^2 + \left[y'\left(\beta_k\right)\right]^2} \cdot \left|t_k - t_{k-1}\right|.$$
(2.1.7)

We observe that the right member of (2.1.7) looks very much like a Riemann sum having as limit

$$\int_a^b \sqrt{x'^2\left(t\right) + y'^2\left(t\right)} \, dt.$$
(2.1.8)

Taking the limit as $n \to \infty$, i.e., when the number of segments of the polygonal line goes to infinity, we can show that $l_n \to l$, on the one hand, and to the integral (2.1.8), on the other hand. Hence, we deduce the formula for the calculus of the arc length

$$l = \int_a^b \sqrt{x'^2(t) + y'^2(t)} \, dt.$$ 

(2.1.9)

**\* Proof.** If the terms under the radical were computed at the *same* point inside the interval $[t_{k-1}, t_k]$, then formula (2.1.7) would really represent a Riemann sum.

Let us denote this radical by

$$R(t,t) = \sqrt{x'^2(t) + y'^2(t)}.$$ 

(2.1.10)

Then

$$l_n = \sum_{k=1}^n R(\alpha_k, \beta_k)(t_k - t_{k-1}).$$ 

(2.1.11)

As $x', y'$ are continuous, they are also uniformly continuous, i.e., for any $\varepsilon > 0$, we can find $\delta = \delta(\varepsilon)$ ($\delta$ depends only on $\varepsilon$!) such that

$$\begin{cases} \left| x'(\alpha_k) - x'(\tau_k) \right| < \varepsilon, \\ \left| y'(\beta_k) - y'(\tau_k) \right| < \varepsilon, \end{cases}$$ 

(2.1.12)

for any $\alpha_k, \beta_k$ with the property $\left| \alpha_k - \tau_k \right| < \delta$, $\left| \beta_k - \tau_k \right| < \delta$.

Therefore

$$\left| R\left(\alpha_k, \beta_k\right) - R\left(\tau_k, \tau_k\right) \right| =$$

$$= \left| \sqrt{x'^2\left(\alpha_k\right) + y'^2\left(\beta_k\right)} - \sqrt{x'^2\left(\tau_k\right) + y'^2\left(\tau_k\right)} \right| =$$

$$= \frac{\left| x'^2\left(\alpha_k\right) - x'^2\left(\tau_k\right) + y'^2\left(\beta_k\right) - y'^2\left(\tau_k\right) \right|}{R\left(\alpha_k, \beta_k\right) + R\left(\tau_k, \tau_k\right)} \leq$$

$$\leq \underbrace{\left| x'\left(\alpha_k\right) - x'\left(\tau_k\right) \right|}_{<\varepsilon} \cdot \underbrace{\frac{\left| x'\left(\alpha_k\right) + x'\left(\tau_k\right) \right|}{R\left(\alpha_k, \beta_k\right) + R\left(\tau_k, \tau_k\right)}}_{<1} + \qquad (2.1.13)$$

$$+ \underbrace{\left| y'\left(\beta_k\right) - y'\left(\tau_k\right) \right|}_{<\varepsilon} \cdot \underbrace{\frac{\left| y'\left(\beta_k\right) + y'\left(\tau_k\right) \right|}{R\left(\alpha_k, \beta_k\right) + R\left(\tau_k, \tau_k\right)}}_{<1},$$

hence $\left| R\left(\alpha_k, \beta_k\right) - R\left(\tau_k, \tau_k\right) \right| < 2\varepsilon$, if $\left| \alpha_k - \tau_k \right| < \delta$ and $\left| \beta_k - \tau_k \right| < \delta$.

If we choose the partition $\Delta$ of norm less than $\delta\left(\varepsilon\right)$ (i.e., $\nu\left(\Delta\right) < \delta\left(\varepsilon\right)$), we can use the following approximation:

$$R\left(\alpha_k, \beta_k\right) = R\left(\tau_k, \tau_k\right) + o\left(\varepsilon\right), \qquad (2.1.14)$$

where $o\left(\varepsilon\right)$ tends to 0 for $\varepsilon \to 0$. Getting back to $l_n$, we have

$$l_n = \underbrace{\sum_{k=1}^{n} R\left(\tau_k, \tau_k\right)\left(t_k - t_{k-1}\right)}_{\sigma_\Delta\left(R(t,t)\right)} +$$

$$+ \underbrace{\sum_{k=1}^{n} \left[ R\left(\alpha_k, \beta_k\right) - R\left(\tau_k, \tau_k\right) \right]\left(t_k - t_{k-1}\right)}_{\to 0}. \qquad (2.1.15)$$

In this expression, $\sigma_\Delta\left(R(t,t)\right)$ is the Riemann sum for the function $\sqrt{x'^2\left(t\right) + y'^2\left(t\right)}$.

The function $R$ is continuous, therefore integrable, whence it follows that

$$\lim_{v(\Delta)\to 0} \sigma_\Delta(R) = \int_a^b \sqrt{x'^2(t) + y'^2(t)}\,dt, \qquad (2.1.16)$$

for any intermediate point $\tau_k$.

Formula (2.1.9) for the calculus of the arc length is therefore proved.

***Particular case***: $y = y(x)$, $x \in [a,b]$. Then $l$ is computed as follows:

$$l = \int_a^b \sqrt{1 + y'^2(t)}\,dt. \qquad (2.1.17)$$

## 2.2. CURVILINEAR INTEGRALS OF FIRST KIND

Let $f \in C^0(\overset{\frown}{AB})$, where the arc $\overset{\frown}{AB}$, defined as

$$\overset{\frown}{AB} : \begin{cases} x = x(t) \\ y = y(t) \end{cases}, \quad t \in [a,b],$$

is ***smooth***, i.e., $x, y \in C^1([a,b])$.

Let us set up the partition $\Delta$ dividing $\overset{\frown}{AB}$ in $n$ subarcs, of ends $A = A_0, A_1, \ldots, A_n = B$. Let $\Delta s_k = l_{\overset{\frown}{A_{k-1}A_k}}$ and consider the intermediate points $M_k(\alpha_k, \beta_k) \in \overset{\frown}{A_{k-1}A_k}$.

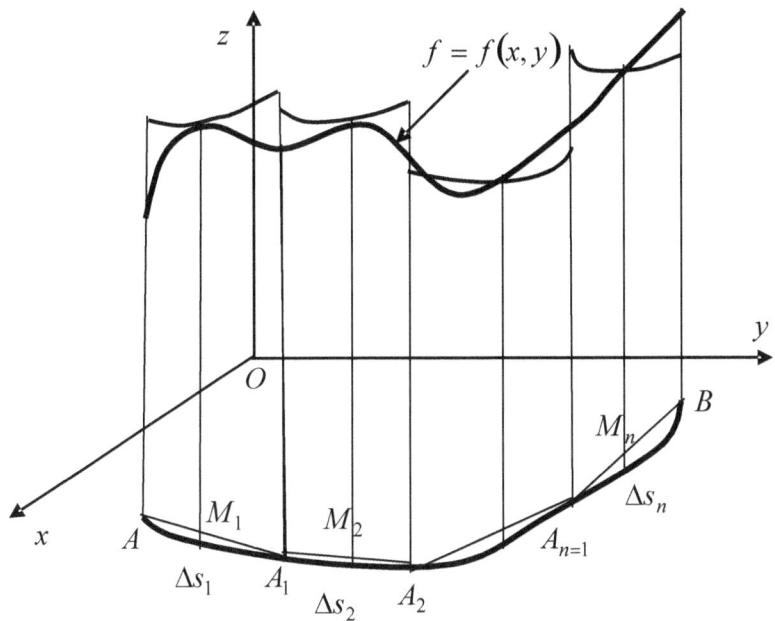

*Figure 2. 2. The geometric representation of the Riemann sum corresponding to the curvilinear integral of the first kind*

We also define the norm of $\Delta$:

$$\nu(\Delta) = \max \left\{ \Delta s_k, \, k = \overline{1,n} \right\}, \tag{2.2.1}$$

setting up the corresponding ***Riemann sum***:

$$\sigma_\Delta(f) = \sum_{k=1}^{n} f(\alpha_k, \beta_k) \cdot \Delta s_k. \tag{2.2.2}$$

**Definition 2.1.** If $\lim\limits_{\nu(\Delta) \to 0} \sigma_\Delta(f, M_k) = I \in \Re$ exists, is

finite and ***it is the same*** for any choice of the intermediate points

$M_k$, then $f$ is called ***integrable*** on $\overset{\frown}{AB}$ and

$$I = \int_{\overset{\frown}{AB}} f(x, y) \, \mathrm{d}s \tag{2.2.3}$$

is *its curvilinear integral of the first kind* on this arc.

We denoted *the element of arc length* by $ds$.

**Definition 2.2. (with ε !)** If there exists a number $I \in \mathfrak{R}$ such that for any $\varepsilon > 0$ one can find $\delta(\varepsilon)$ with the property that $\left| \sigma_\Delta(f, M_k) - I \right| < \varepsilon$, for any partition $\Delta$ of norm $v(\Delta) < \delta$ and for any choice of the intermediate points $M_k$, then $f$ is called *integrable on* $\overset{\frown}{AB}$ and its *first kind curvilinear integral* on this arc is expressed by the relation (2.2.3).

*Remark.* The curvilinear integrals are also called *line integrals* or *contour integrals*. However, here we shall use only the denomination of *curvilinear integral*, because we find that this is a more suggestive and even a more general term, including integrals taken on right lines ("line") or on closed curves ("contour").

*Notation:* If the arc represents a closed curve $C \subset \mathfrak{R}^2$, then we use the following notation for the curvilinear integral:

$$\oint_C f(x, y) ds.$$

## 2.2.1. THE COMPUTATION OF THE CURVILINEAR INTEGRAL OF THE FIRST KIND

If $\overset{\frown}{AB}: \begin{cases} x = x(t) \\ y = y(t) \end{cases}$, $t \in [a, b]$, then the element of arc length is expressed as follows:

$$\mathrm{d}s^2 = \mathrm{d}x^2 + \mathrm{d}y^2 = \left[ x'^2(t) + y'^2(t) \right] \mathrm{d}t . \qquad (2.2.4)$$

This yields

$$\mathrm{d}s = \sqrt{x'^2(t) + y'^2(t)}\,\mathrm{d}t , \qquad (2.2.5)$$

and $f(x,y) = f(x(t), y(t))$.

We can prove the following natural formula for the calculation of the first kind curvilinear integral:

$$\int_{\overset{\frown}{AB}} f(x,y)\,\mathrm{d}s = \int_a^b f(x(t), y(t))\sqrt{x'^2(t) + y'^2(t)}\,\mathrm{d}t . \quad (2.2.6)$$

The integral in the right member is a Riemann integral.

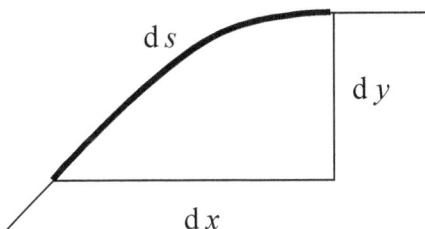

*Figure 2. 3. The element of arc length*

If $\overset{\frown}{AB}: y = y(x),\ x \in [a,b]$, then

$$\int_{\overset{\frown}{AB}} f(x,y)\,\mathrm{d}s = \int_a^b f(x, y(x))\sqrt{1 + y'^2(x)}\,\mathrm{d}x . \qquad (2.2.7)$$

*Examples*

1. Find the length of the arc $\overset{\frown}{AB}: y = \mathrm{ch}\,x,\ 0 \le x \le \ln 2$ (**the catenary**) and compute $\int_{\overset{\frown}{AB}} y\,\mathrm{d}s$.

**Solution.** To compute the length of the catenary, we apply formula (2.1.17); to compute the involved curvilinear integral, we use formula (2.2.7).

We get, step by step,

$$l_{\overset{\frown}{AB}} = \int_0^{\ln 2} \sqrt{1 + y'^2(x)}\, dx = \int_0^{\ln 2} \sqrt{1 + \text{sh}^2 x}\, dx = \int_0^{\ln 2} \text{ch}\, x\, dx =$$

$$= \text{sh}\, x\Big|_0^{\ln 2} = \frac{e^{\ln 2} - e^{-\ln 2}}{2} = \frac{2 - \frac{1}{2}}{2} = \frac{3}{4} \Rightarrow \boxed{l_{\overset{\frown}{AB}} = \frac{3}{4}}, \tag{2.2.8}$$

$$\int_{\overset{\frown}{AB}} y\, ds = \int_0^{\ln 2} \text{ch}\, x \cdot \sqrt{1 + \text{sh}^2 x}\, dx =$$

$$= \int_0^{\ln 2} \text{ch}^2 x\, dx = \frac{1}{2} \int_0^{\ln 2} (1 + \text{ch}\, 2x)\, dx =$$

$$= \frac{1}{2}\ln 2 + \frac{\text{sh}\, 2x}{4}\Big|_{x=0}^{x=\ln 2} = \frac{\ln 2}{2} + \frac{1}{4} \cdot \frac{1}{2}\left(e^{2\ln 2} - e^{-2\ln 2}\right) = \tag{2.2.9}$$

$$= \frac{\ln 2}{2} + \frac{1}{8}\left(4 - \frac{1}{4}\right) \Rightarrow \boxed{\int_{\overset{\frown}{AB}} y\, ds = \frac{15}{32} + \frac{\ln 2}{2}}.$$

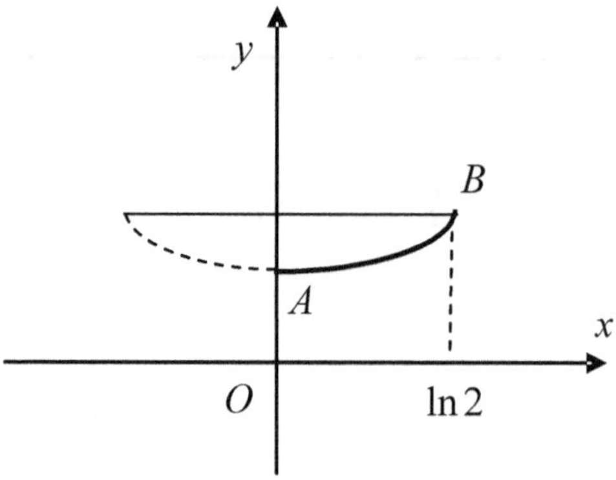

*Figure 2.4. The catenary*

90

2. Find the length of the arc $\overset{\frown}{AB}:\begin{cases} x = \cos t \\ y = \sin t \end{cases}, t \in [0, \pi].$

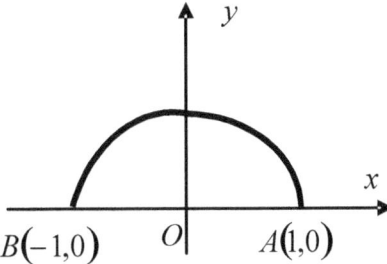

*Figure 2. 5. The arc defined in the second example*

**Solution.** We apply formula (2.2.6). We have

$$l_{\overset{\frown}{AB}} = \int_0^\pi \sqrt{x'^2(t) + y'^2(t)}\, dt =$$

$$= \int_0^\pi \sqrt{(-\sin t)^2 + (\cos t)^2}\, dt = \int_0^\pi dt,$$

(2.2.10)

therefore

$$l_{\overset{\frown}{AB}} = \pi.$$

(2.2.11)

## 2.2.2. PROPERTIES OF THE FIRST KIND CURVILINEAR INTEGRAL

We shall enounce several important properties of the curvilinear integral of the first kind, which can be straightforwardly deduced from its definition and from the form of the Riemann sums.

We specify that the **direct sense** on an arc or a closed curve is **counterclockwise.**

**1) INDEPENDENCE OF SENSE ON THE ARC.** The curvilinear integral of the first kind is **independent** of the sense on the arc, i.e.

$$\int_{\overset{\frown}{AB}} f(x,y)\,\mathrm{d}s = \int_{\overset{\frown}{BA}} f(x,y)\,\mathrm{d}s. \qquad (2.2.12)$$

**2) ADDITIVITY WITH RESPECT TO THE DOMAIN OF INTEGRATION.** If $f$ is integrable on the arcs $\overset{\frown}{AC}, \overset{\frown}{CB}$, then it is also integrable on their union and

$$\int_{\overset{\frown}{AB}} f(x,y)\,\mathrm{d}s = \int_{\overset{\frown}{AC}} f(x,y)\,\mathrm{d}s + \int_{\overset{\frown}{CB}} f(x,y)\,\mathrm{d}s. \qquad (2.2.13)$$

*Remark.* In the figure 2.6 below, as well as in formula (2.2.13), the two arcs have a coincident end.

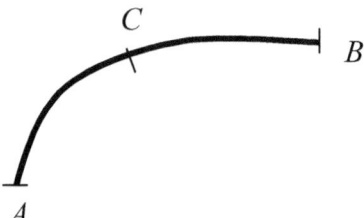

*Figure 2. 6. The position of the arcs used to express the property of additivity*

However, the property of additivity is valid for any arcs, provided they have no arc portions in common.

**3) LINEARITY.** The first kind curvilinear integral is linear. If $f, g$ are integrable on $\overset{\frown}{AB}$, then

$$\int_{\overset{\frown}{AB}} (\alpha f + \beta g)(x,y)\,\mathrm{d}s = \alpha \int_{\overset{\frown}{AB}} f(x,y)\,\mathrm{d}s + \beta \int_{\overset{\frown}{AB}} g(x,y)\,\mathrm{d}s, \qquad (2.2.14)$$

for any $\alpha, \beta \in \mathfrak{R}$.

## 2.2.3. APPLICATIONS IN GEOMETRY AND MECHANICS

**1.** $l_{\overset{\frown}{AB}} = \int\limits_{\overset{\frown}{AB}} \mathrm{d}s$ is **the length of the arc** $\overset{\frown}{AB}$. $\qquad$ (2.2.15)

**2.** If $f(x,y) = \rho(x,y) > 0$ is the linear density of $\overset{\frown}{AB}$,

then

$$m_{\overset{\frown}{AB}} = \int\limits_{\overset{\frown}{AB}} \rho(x,y)\,\mathrm{d}s \qquad (2.2.16)$$

is **the mass of the arc**.

**3. The center of mass of the arc** $\overset{\frown}{AB}$ is given by its coordinates:

$$\overline{x} = \frac{\int\limits_{\overset{\frown}{AB}} x\rho(x,y)\,\mathrm{d}s}{\underbrace{\int\limits_{\overset{\frown}{AB}} \rho(x,y)\,\mathrm{d}s}_{m_{\overset{\frown}{AB}}}}, \quad \overline{y} = \frac{\int\limits_{\overset{\frown}{AB}} y\rho(x,y)\,\mathrm{d}s}{\underbrace{\int\limits_{\overset{\frown}{AB}} \rho(x,y)\,\mathrm{d}s}_{m_{\overset{\frown}{AB}}}}. \qquad (2.2.17)$$

**4.** To get **the geometric center of mass,** we put $\rho = 1$ in (2.2.17):

$$\overline{\overline{x}} = \frac{\int\limits_{\overset{\frown}{AB}} x\,\mathrm{d}s}{\underbrace{\int\limits_{\overset{\frown}{AB}} \mathrm{d}s}_{l_{\overset{\frown}{AB}}}}, \quad \overline{\overline{y}} = \frac{\int\limits_{\overset{\frown}{AB}} y\,\mathrm{d}s}{\underbrace{\int\limits_{\overset{\frown}{AB}} \mathrm{d}s}_{l_{\overset{\frown}{AB}}}}. \qquad (2.2.18)$$

93

## IMPORTANT!

If $\overset{\frown}{AB}$ is homogeneous, then its linear density is constant, i.e., $\rho = c = const$. It follows that

$$\overline{x} = \frac{\displaystyle\int_{\overset{\frown}{AB}} xc\,ds}{\displaystyle\int_{\overset{\frown}{AB}} c\,ds} = \frac{\displaystyle\&\int_{\overset{\frown}{AB}} x\,ds}{\displaystyle\&\int_{\overset{\frown}{AB}} ds} = \overset{=}{x}, \quad \overline{y} = \overset{=}{y} \text{ respectively.} \quad (2.2.19)$$

**CONCLUSION.** *The center of mass of a homogeneous arc coincides with its geometric center of mass.*

*Example.* Compute the coordinates of the center of mass of the homogeneous catenary $\overset{\frown}{AB} : y = \text{ch}\, x$, $x \in \left[0, \ln 2\right]$.

**Solution.** We must compute

$$\overline{x} = \frac{\displaystyle\int_{\overset{\frown}{AB}} x\,ds}{\displaystyle\int_{\overset{\frown}{AB}} ds}, \quad \overline{y} = \frac{\displaystyle\int_{\overset{\frown}{AB}} y\,ds}{\displaystyle\int_{\overset{\frown}{AB}} ds}. \quad (2.2.20)$$

We already computed $l_{\overset{\frown}{AB}} = \int_{\overset{\frown}{AB}} ds = \dfrac{3}{4}$ in the previous examples. We also computed $\int_{\overset{\frown}{AB}} y\,ds = \dfrac{15}{32} + \dfrac{\ln 2}{2}$ and we still have to compute

$$\int_{\overset{\frown}{AB}} x\,ds = \int_0^{\ln 2} x\sqrt{1+\operatorname{sh}^2 x}\;dx = \int_0^{\ln 2} x\operatorname{ch} x\,dx = x\operatorname{sh} x\Big|_0^{\ln 2} -$$

$$-\int_0^{\ln 2} \operatorname{sh} x\,dx = \ln 2 \cdot \frac{2-\dfrac{1}{2}}{2} - \operatorname{ch} x\Big|_0^{\ln 2} = \tag{2.2.21}$$

$$= \frac{3}{4}\ln 2 - \frac{e^{\ln 2}+e^{-\ln 2}}{2} + 1 =$$

$$= \frac{3}{4}\ln 2 - \frac{5}{4} + 1 = \frac{3}{4}\ln 2 - \frac{1}{4}.$$

It follows that

$$\begin{aligned}
\overline{x} &= \frac{4}{3}\left(\frac{3}{4}\ln 2 - \frac{1}{4}\right) \\
\overline{y} &= \frac{4}{3}\left(\frac{1}{2}\ln 2 + \frac{15}{32}\right)
\end{aligned}
\;\;\Rightarrow\;\;
\begin{cases}
\overline{x} = \ln 2 - \dfrac{1}{3} \\[2mm]
\overline{y} = \dfrac{2}{3}\ln 2 + \dfrac{5}{8}.
\end{cases}
\tag{2.2.22}$$

## 2.3. FIRST KIND CURVILINEAR INTEGRALS ON SKEW CURVES

Consider the arc $\overset{\frown}{AB} \subset \Re^3$, of equations

$$\overset{\frown}{AB}:\begin{cases} x = x(t) \\ y = y(t), \quad t \in [a,b] \subset \Re. \\ z = z(t) \end{cases}$$

We assume that $\overset{\frown}{AB}$ is smooth, which means that $x, y, z$ are of class $C^1$ on $[a,b]$. Using the same method as in the previous paragraph, we can show that the arc length is given by

95

$$l_{\overset{\frown}{AB}} = \int_a^b \sqrt{x'^2(t) + y'^2(t) + z'^2(t)}\, dt. \qquad (2.3.1)$$

If $f : \overset{\frown}{AB} \to \mathfrak{R}$, we define $\displaystyle\int_{\overset{\frown}{AB}} f(x,y,z)ds$ as in the case

of the plane curves, with the corresponding Riemann sum.

*Notation*: If the arc represents a closed curve $C \subset \mathfrak{R}^3$, then we use the following notation for the curvilinear integral:

$$\oint_C f(x,y,z)\,ds.$$

## 2.3.1. THE CALCULATION OF CURVILINEAR INTEGRALS OF THE FIRST KIND ON SKEW CURVES

We proceed as in the paragraph 2.2.

The element of arc length $ds$ is here

$$ds^2 = dx^2 + dy^2 + dz^2 \Rightarrow$$
$$ds = \sqrt{x'^2(t) + y'^2(t) + z'^2(t)}\, dt, \qquad (2.3.2)$$

and it approximates the diagonal of the infinitesimal parallelepiped of edges $dx$, $dy$, $dz$ from the figure 2.7.

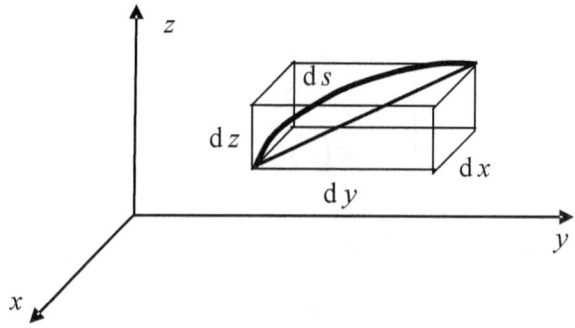

*Figure 2. 7. The element of arc length for a skew curve*

**2.** For $f(x, y, z) = f(x(t), y(t), z(t))$, we compute the integral as follows

$$\int\limits_{\overset{\frown}{AB}} f(x, y, z)\, ds =$$

$$= \int\limits_a^b f(x(t), y(t), z(t)) \sqrt{x'^2(t) + y'^2(t) + z'^2(t)}\, dt. \qquad (2.3.3)$$

The properties of the curvilinear integral of the first kind on twisted curves are the same as in the plane case, but accordingly adapted to the curves in space.

### 2.3.2. APPLICATIONS IN GEOMETRY AND MECHANICS

**1. *The length of an arc*** is

$$l_{\overset{\frown}{AB}} = \int\limits_{\overset{\frown}{AB}} ds. \qquad (2.3.4)$$

By putting $f = 1$ in the definition of the Riemann sum, it follows that

$$\sigma_\Delta(f) = \Delta s_1 + \Delta s_2 + \ldots + \Delta s_n = l_{\overset{\frown}{AB}},$$

for any partition $\Delta$.

**2. *The mass of an arc*:** if $\rho = \rho(x, y, z) > 0$ is **the linear density of the arc**, it follows that

$$m_{\overset{\frown}{AB}} = \int\limits_{\overset{\frown}{AB}} \rho(x, y, z)\, ds. \qquad (2.3.5)$$

**3. *The center of mass*** of an arc has the coordinates

$$\overline{x} = \frac{\int\limits_{\overset{\frown}{AB}} x\rho(x,y)ds}{m_{\overset{\frown}{AB}}}, \quad \overline{y} = \frac{\int\limits_{\overset{\frown}{AB}} y\rho(x,y)ds}{m_{\overset{\frown}{AB}}},$$

$$\overline{z} = \frac{\int\limits_{\overset{\frown}{AB}} z\rho(x,y)ds}{m_{\overset{\frown}{AB}}}.$$

$(2.3.6)$

**4. The geometric center of mass** of an arc $(\rho=1)$ has the coordinates

$$\overline{x} = \frac{\int\limits_{\overset{\frown}{AB}} x\,ds}{l_{\overset{\frown}{AB}}}, \quad \overline{y} = \frac{\int\limits_{\overset{\frown}{AB}} y\,ds}{l_{\overset{\frown}{AB}}}, \quad \overline{z} = \frac{\int\limits_{\overset{\frown}{AB}} z\,ds}{l_{\overset{\frown}{AB}}}. \qquad (2.3.7)$$

As in the case of plane curves, *the center of mass of a homogeneous arc coincides with its geometric center of mass.*

*Example.* Compute the mass of the arc of equations

$$\overset{\frown}{AB}:\begin{cases} x = t \\ y = t^2/2, \quad t \in [0,1], \\ z = t^3/3 \end{cases}$$

if its linear density is $\rho(x,y,z) = x + 6z$.

**Solution.** According to application 3,

$$m_{\overset{\frown}{AB}} = \int\limits_{\overset{\frown}{AB}} \rho(x,y,z)ds. \qquad (2.3.8)$$

We have

$$\rho(x(t),y(t),z(t)) = t + 6\cdot\frac{t^3}{6} = t + 2t^3,$$

$$ds = \sqrt{x'^2(t)+y'^2(t)+z'^2(t)}\,dt = \sqrt{1^2 + t^2 + (t^2)^2}\,dt. \qquad (2.3.9)$$

Therefore

$$m_{\overset{\frown}{AB}} = \int_0^1 \left(t + 2t^3\right)\sqrt{1 + t^2 + t^4}\, dt =$$

$$= \frac{1}{2}\int_0^1 \sqrt{1 + t^2 + t^4}\, d\left(1 + t^2 + t^4\right) = \qquad (2.3.10)$$

$$= \frac{1}{2} \cdot \frac{\left(1 + t^2 + t^4\right)^{\frac{3}{2}}}{\frac{3}{2}}\Bigg|_0^1 = \frac{\left(1 + t^2 + t^4\right)^{\frac{3}{2}}}{3}\Bigg|_0^1 = \frac{3^{\frac{3}{2}}}{3} - \frac{1}{3};$$

the mass of the arc is

$$\boxed{m_{\overset{\frown}{AB}} = \sqrt{3} - \frac{1}{3}.} \qquad (2.3.11)$$

## 2.4. SECOND KIND CURVILINEAR INTEGRALS

Let $\overset{\frown}{AB} \subset \Re^2$ be a smooth arc. Consider the partition $\Delta = \{A = M_0, M_1, \ldots, M_n = B\}$. We project $\overset{\frown}{M_{k-1}M_k}$, $k = \overline{1,n}$, on the $Ox$, $Oy$ axes and we denote

$$\Delta x_k = x_k - x_{k-1},$$
$$\Delta y_k = y_k - y_{k-1}, \quad k = \overline{1,n}.$$

On each arc of the partition, we consider an arbitrary point $S_k\left(\alpha_k, \beta_k\right) \in \overset{\frown}{M_{k-1}M_k}$, $k = \overline{1,n}$.

Let now $P, Q : \overset{\frown}{AB} \to \Re$ be two functions defined on the arc. We set up the Riemann sum

99

$$\sigma_{\Delta}(P, Q, S_k) =$$

$$= \sum_{k=1}^{n} \left[ P(\alpha_k, \beta_k) \Delta x_k + Q(\alpha_k, \beta_k) \Delta y_k \right]. \qquad (2.4.1)$$

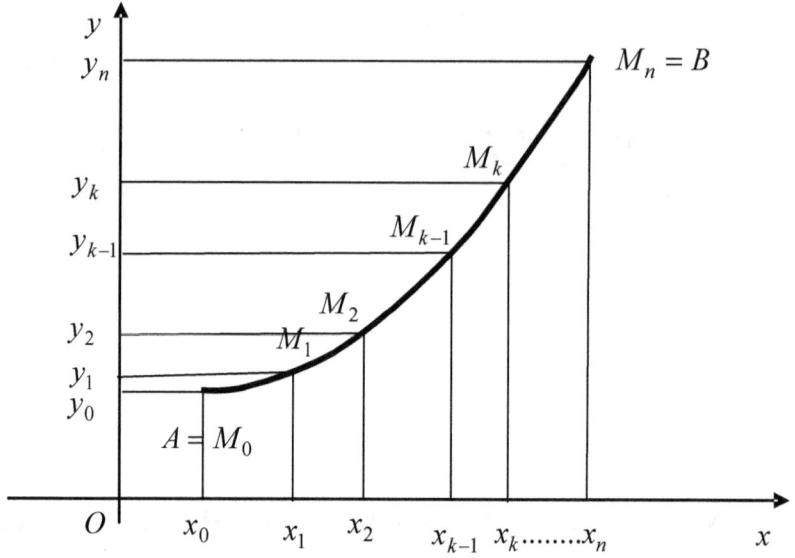

*Figure 2. 8. The smooth arc for the definition of the curvilinear integral of the second kind*

If $\lim\limits_{\substack{\max|\Delta x_k| \to 0 \\ \max|\Delta y_k| \to 0}} \sigma_{\Delta}(P, Q, S_k) = I$ for any choice of the

intermediate points, then $I$ is **the curvilinear integral of the second kind** and we write

$$I = \int_{\overset{\frown}{AB}} P(x, y) dx + Q(x, y) dy. \qquad (2.4.2)$$

As in the case of the first kind curvilinear integral, if the integral is defined on a closed curve $C$, we use the following notation: $I = \oint_{C} P(x, y) dx + Q(x, y) dy$.

## 2.4.1. PROPERTIES OF THE CURVILINEAR INTEGRAL OF THE SECOND KIND

These properties are deduced directly from the definition of the Riemann sum, as in the case of the first kind curvilinear integral.

**1.** The properties of *linearity* and *additivity* with *respect to the domain of integration* hold also true for the second kind curvilinear integral.

**2.**

$$\int\limits_{\overset{\frown}{AB}} P(x,y)dx + Q(x,y)dy = -\int\limits_{\overset{\frown}{BA}} P(x,y)dx + Q(x,y)dy, \tag{2.4.3}$$

therefore, unlike the first kind, the second kind curvilinear integral changes its sign if one changes the sense on the arc.

**3.**

$$\int\limits_{\overset{\frown}{AB}} P(x,y)dx + Q(x,y)dy = \int\limits_{\overset{\frown}{AB}} P(x,y)dx + \int\limits_{\overset{\frown}{AB}} Q(x,y)dy \cdot \tag{2.4.4}$$

## 2.4.2. THE CALCULATION OF THE CURVILINEAR INTEGRALS OF THE SECOND KIND

If $\overset{\frown}{AB} : \begin{cases} x = x(t) \\ y = y(t) \end{cases}, t \in [a,b]$, then

$$dx = x'(t)dt, \quad dy = y'(t)dt,$$

and the integral becomes

$$\int_{\overset{\frown}{AB}} P(x,y)\mathrm{d}x + Q(x,y)\mathrm{d}y =$$

$$= \int_a^b \left[ P(x(t),y(t))x'(t) + Q(x(t),y(t))y'(t) \right] \mathrm{d}t. \qquad (2.4.5)$$

### 2.4.3. PHYSICAL INTERPRETATION OF THE SECOND KIND CURVILINEAR INTEGRAL

♣   If $\mathbf{F}(P,Q)$ is a force of components $P$, $Q$, then

$$\mathcal{L}(\mathbf{F}) = \int_{\overset{\frown}{AB}} P(x,y)\mathrm{d}x + Q(x,y)\mathrm{d}y \qquad (2.4.6)$$

is the **mechanical work** done by $\mathbf{F}$ along the arc $\overset{\frown}{AB}$.

♣   If $\mathbf{V}(P,Q)$ is a plane vector field, then

$$c = \int_{\overset{\frown}{AB}} P(x,y)\mathrm{d}x + Q(x,y)\mathrm{d}y \qquad (2.4.7)$$

is its **circulation** along the arc $\overset{\frown}{AB}$.

### 2.4.4. THE PATH INDEPENDENCE OF THE SECOND KIND CURVILINEAR INTEGRAL

For a better understanding, let us give several examples.

*Example*. Compute the mechanical work done by the force

$\mathbf{F}$ of components $\mathbf{F}(x+y, x-y)$ along an arc $\overset{\frown}{OM}$, passing

through the points $O(0,0)$ and $M(\pi,\pi)$, in cases *a)*, *b)*, *c)* as specified below.

**Solution.**

$$\mathscr{L}(\mathbf{F}) = \int\limits_{\overset{\frown}{OM}} (x+y)\mathrm{d}x + (x-y)\mathrm{d}y. \qquad (2.4.8)$$

*a)* The path joining the points $O$ and $M$ is the segment $\overline{OM}$
( figure 2.9).

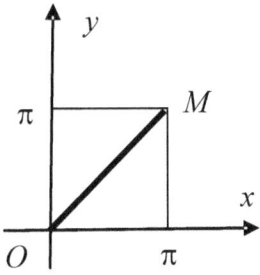

*Figure 2. 9*

We have

$$\overline{OM} : y = x,\ \mathrm{d}y = \mathrm{d}x \quad \Rightarrow \quad I = \int\limits_{0}^{\pi} 2x\mathrm{d}x \quad \Rightarrow \quad \boxed{I = \pi^2}. \qquad (2.4.9)$$

*b)* The path $\overset{\frown}{OM}$ is now the parabola $\overset{\frown}{OM} : y = \dfrac{x^2}{\pi}$ (figure

2.10).

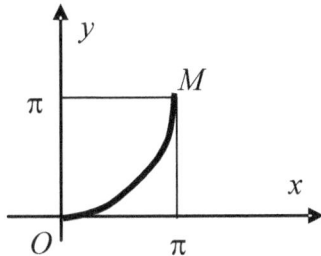

*Figure 2. 10*

Then

$$dy = \frac{2x}{\pi}dx \quad \Rightarrow \quad I = \int_0^\pi \left[ x + \frac{x^2}{\pi} + \frac{2x}{\pi}\left(x - \frac{x^2}{\pi}\right) \right] dx, \quad (2.4.10)$$

therefore

$$I = \frac{\pi^2}{2} + \frac{x^3}{\pi}\Big|_0^\pi - \frac{2x^4}{4\pi^2}\Big|_0^\pi \quad \Rightarrow \quad \boxed{I = \pi^2}. \qquad (2.4.11)$$

c) The path $\overset{\frown}{OM}$ is the segment union $\overline{OA} \cup \overline{AM}$ ( figure 2.11).

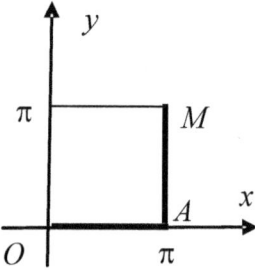

Figure 2. 11

We have

$$I = \int_{\overline{OA}} (x + y)dx + (x - y)\underset{=0}{\underbrace{dy}} + \int_{\overline{AM}} (x + y)\underset{=0}{\underbrace{dx}} + (x - y)dy =$$

$$= \int_0^\pi x\,dx + \int_0^\pi (\pi - y)dy \Rightarrow \boxed{I = \pi^2}. \qquad (2.4.12)$$

We see that ***the integral has the same value, no matter the path.*** This fact leads to the natural question: ***does this always occur?***

The answer is ***negative***, as emphasized by the following counterexample.

*Counter-example.* Compute the mechanical work done by the force $\mathbf{F}(y, 2x)$ along an arc $\overset{\frown}{OM}$ which passes through the points $O(0,0)$ and $M(\pi, \pi)$, considering the paths from the previous example.

**Solution.** The figures from the previous example are, obviously, the same for all three cases. We have

**a)** $\overline{OM}: y = x,\ \mathrm{d}y = \mathrm{d}x \quad \Rightarrow \quad I = \int_0^\pi 3x\,\mathrm{d}x = \dfrac{3x^2}{2}\Big|_0^\pi \rightarrow$ 

$$I = \frac{3\pi^2}{2}.$$

(2.4.13)

**b)** $\overset{\frown}{OM}: y = \dfrac{x^2}{\pi},\quad \mathrm{d}y = \dfrac{2x}{\pi}\,\mathrm{d}x \quad \Rightarrow$

(2.4.14)

$$I = \int_0^\pi \left(\frac{x^2}{\pi} + 2x \cdot \frac{2x}{\pi}\right)\mathrm{d}x = \int_0^\pi \frac{5x^2}{\pi}\,\mathrm{d}x = \frac{5x^3}{3\pi}\Big|_0^\pi;$$

it follows that

$$I = \frac{5\pi^2}{3}.$$

(2.4.15)

**c)** $\overline{OA} \cup \overline{AM}:$

(2.4.16)

$$I = \underbrace{\int\limits_{\overline{OA}} y\,\mathrm{d}x}_{=0} + 2x\,\underbrace{\mathrm{d}y}_{=0} + \underbrace{\int\limits_{\overline{AM}} y\,\mathrm{d}x}_{=0} + 2x\,\mathrm{d}y = \int_0^\pi 2\pi\,\mathrm{d}y,$$

whence

$$\boxed{I = 2\pi^2.}$$

(2.4.17)

The following question naturally arises:

## WHEN IS THE CURVILINEAR INTEGRAL PATH INDEPENDENT AND WHEN IT IS NOT?

In order to answer this question, we shall firstly prove

**Theorem 2.1.** *Let* $P, Q \in C^0(D)$*, where* $D \subset \mathfrak{R}^2$ *is a bounded and closed plane domain. Suppose that* $\int_\Gamma P(x, y)dx + Q(x, y)dy$*,* $\Gamma \subset D$*, is path independent. Then one can find a function* $F \in C^1(D)$ *such that*

$$dF(x, y) = P(x, y)dx + Q(x, y)dy. \qquad (2.4.18)$$

* **Proof.** Let $M_0(x_0, y_0) \in D$ be arbitrary (but fixed) and take a point $M(x, y) \in D$, varying in D. Let C be an arbitrary arc, entirely included in D, joining these two points. As the curvilinear integral is path independent, we can define, by hypothesis, the function

$$F(x, y) = \int_C P(x, y)dx + Q(x, y)dy. \qquad (2.4.19)$$

Take now the point $M'(x + h, y)$. The segment $\overline{MM'}$ is parallel to $Ox$ (figure 2.12).

Let $\Gamma = C \cup \overline{MM'}$. According to the definition of $F$,

$$F(x + h, y) = \int_\Gamma P(x, y)dx + Q(x, y)dy =$$

$$= \underbrace{\int_C P(x, y)dx + Q(x, y)dy}_{F(x,y)} +$$

$$+ \int_{\overline{MM'}} P(x, y)dx + Q(x, y)\underbrace{dy}_{=0}.$$

(2.4.20)

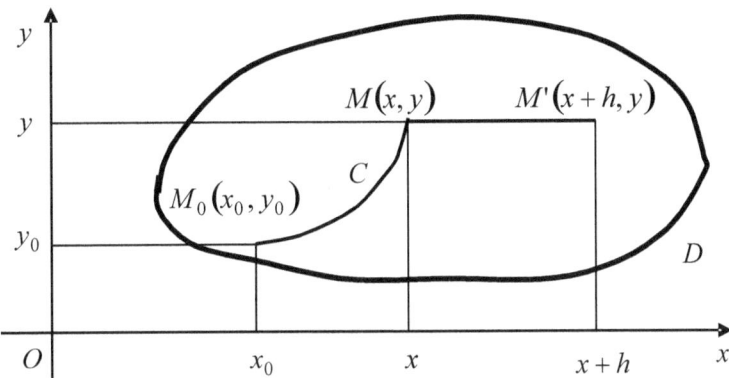

*Figure 2. 12. The proof of path independence of the second kind curvilinear integral*

Therefore

$$F(x+h, y) - F(x, y) = \int\limits_{\overline{MM'}} P(x, y)\mathrm{d}x = \int\limits_{x}^{x+h} P(t, y)\mathrm{d}t. \quad (2.4.21)$$

We use Lagrange's theorem in the last integral:

$$\int\limits_{x}^{x+h} P(t, y)\mathrm{d}t = P(\alpha, y)\cdot(x+h-x) = P(\alpha, y)\cdot h, \quad (2.4.22)$$

$$\alpha \in (x, x+h).$$

The relation (2.4.21) becomes

$$\frac{F(x+h, y) - F(x, y)}{h} = P(\alpha, y). \quad (2.4.23)$$

But $P \in C^0(D)$, therefore $\lim\limits_{h\to 0} P(\alpha, y) = P(x, y)$.

It follows that the limit

$$\lim\limits_{h\to 0} \frac{F(x+h, y) - F(x, y)}{h} \equiv \frac{\partial F}{\partial x}(x, y)$$

exists as well and it is finite. We thus proved that

$$\frac{\partial F}{\partial x}(x,y) = P(x,y), \ (x,y) \in D.$$

Similarly, we can prove that $\frac{\partial F}{\partial y}(x,y) = Q(x,y)$ in $D$.

Due to the properties of the differential of a function, it follows that

$$dF(x,y) = P(x,y)dx + Q(x,y)dy. \tag{2.4.24}$$

Conversely, if $\ P(x,y)dx + Q(x,y)dy = 0 \quad$ is an exact/total differential equation, then the value of the integral

$$\int_{\overset{\frown}{AB}} P(x,y)dx + Q(x,y)dy \tag{2.4.25}$$

does not depend on the path $\overset{\frown}{AB}$, but only on its ends. Indeed, let

$$P(x,y)dx + Q(x,y)dy = dF(x,y) \tag{2.4.26}$$

and let

$$\begin{cases} x = x(t), \\ y = y(t), \end{cases} \quad t \in [a,b], \tag{2.4.27}$$

be the parametric equations of $\overset{\frown}{AB}$. Then, the coordinates of its ends are $A(x(a), y(a)), B(x(b), y(b))$. We compute the integral:

$$\int_{\overset{\frown}{AB}} P(x,y)dx + Q(x,y)dy =$$

$$= \int_a^b \left( P(x,y)\frac{dx}{dt} + Q(x,y)\frac{dy}{dt} \right) dt = \tag{2.4.28}$$

$$= \int_a^b \left( \frac{\partial F}{\partial x}\frac{dx}{dt} + \frac{\partial F}{\partial y}\frac{dy}{dt} \right) dt = \int_a^b \frac{dF}{dt} dt,$$

where we applied the chain rule to the function $F\left(x\left(t\right),y\left(t\right)\right)$ (see [4,10]):

$$\frac{\mathrm{d}F}{\mathrm{d}t} = \frac{\partial F}{\partial x}\frac{\mathrm{d}x}{\mathrm{d}t} + \frac{\partial F}{\partial y}\frac{\mathrm{d}y}{\mathrm{d}t} = P\left(x,y\right)\frac{\mathrm{d}x}{\mathrm{d}t} + Q\left(x,y\right)\frac{\mathrm{d}y}{\mathrm{d}t}. \qquad (2.4.29)$$

From (2.4.28), (2.4.29) we deduce

$$\int_{\overset{\frown}{AB}} P\left(x,y\right)\mathrm{d}x + Q\left(x,y\right)\mathrm{d}y = F\left(B\right) - F\left(A\right), \qquad (2.4.30)$$

no matter what are the functions (2.4.27) , i.e., no matter what is the path joining the points $A$ and $B$.

The above-mentioned remarks and theorem form a basis for the following theorem, which we enounce without proof.

**Theorem 2. 2.** (THE PATH INDEPENDENCE OF THE SECOND KIND CURVILINEAR INTEGRAL). *Let* $P,Q \in C^{1}\left(D\right)$, *where* $D \subset \mathfrak{R}^{2}$ *is closed and bounded. The necessary and sufficient condition for the path independence of the integral*

$$\int_{\overset{\frown}{AB}} P\left(x,y\right)\mathrm{d}x + Q\left(x,y\right)\mathrm{d}y, \ \overset{\frown}{AB} \subset D, \text{ is that}$$

$$\frac{\partial P}{\partial y}\left(x,y\right) = \frac{\partial Q}{\partial x}\left(x,y\right), \quad \left(x,y\right) \in D. \qquad (2.4.31)$$

*Remark.* Let $\Gamma \subset D$ be a closed curve and $A,B \in \Gamma$ (figure 2.13). Then

$$\Gamma = \overset{\frown}{AnB} \cup \overset{\frown}{BmA}. \qquad (2.4.32)$$

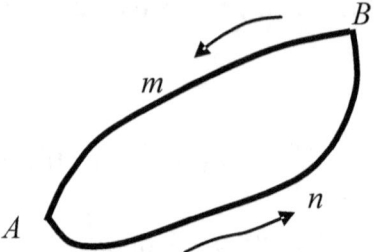

*Figure 2. 13. The curvilinear integral of the second kind on a closed curve*

Suppose $\dfrac{\partial P}{\partial y} = \dfrac{\partial Q}{\partial x}$ in $D$. It follows that

$$
\begin{aligned}
I &= \oint_C P\left(x, y\right) \mathrm{d}\,x + Q\left(x, y\right) \mathrm{d}\,y \\
&= \int_{\overset{\frown}{AnB}} P\left(x, y\right) \mathrm{d}\,x + Q\left(x, y\right) \mathrm{d}\,y + \\
&+ \int_{\overset{\frown}{BmA}} P\left(x, y\right) \mathrm{d}\,x + Q\left(x, y\right) \mathrm{d}\,y = \\
&= \int_{\overset{\frown}{AnB}} P\left(x, y\right) \mathrm{d}\,x + Q\left(x, y\right) \mathrm{d}\,y - \\
&- \int_{\overset{\frown}{AmB}} P\left(x, y\right) \mathrm{d}\,x + Q\left(x, y\right) \mathrm{d}\,y = 0,
\end{aligned}
$$
(2.4.33)

according to the theorem 2.2. Therefore

**If** $\dfrac{\partial P}{\partial y} = \dfrac{\partial Q}{\partial x}$, **then** $\oint_\Gamma P\left(x, y\right) \mathrm{d}\,x + Q\left(x, y\right) \mathrm{d}\,y$ **on any**

**closed curve** $\Gamma \subset D$.

**APPLICATION.** Let us reconsider the previous examples.

**1.** By computing the integral

$$
I = \int_{\overset{\frown}{OM}} \underbrace{\left(x + y\right)}_{P} \mathrm{d}x + \underbrace{\left(x - y\right)}_{Q} \mathrm{d}y,
$$
(2.4.34)

we obtained the same value for different paths $\overset{\frown}{OM}$ .

Let us compute the partial derivatives $\dfrac{\partial P}{\partial y}, \dfrac{\partial Q}{\partial x}$. We have

$$\frac{\partial P}{\partial y} = \frac{\partial}{\partial y}(x+y) = 1,$$

$$\frac{\partial Q}{\partial x} = \frac{\partial}{\partial x}(x-y) = 1,$$

(2.4.35)

therefore $\dfrac{\partial P}{\partial y} = \dfrac{\partial Q}{\partial x}$. According to the theorem 2.2, the integral is

**path independent.**

For any path joining the points $O$ and $M$, we always obtain the same value of the integral.

In this case, we say that **the force F is conservative.**

**2.** By computing the integral

$$I = \int_{\overset{\frown}{OM}} \underset{P}{y}\,dx + \underset{Q}{2x}\,dy,$$

(2.4.36)

we obtain different values for different paths $\overset{\frown}{OM}$.

The partial derivatives $\dfrac{\partial P}{\partial y}, \dfrac{\partial Q}{\partial x}$ are

$$\frac{\partial P}{\partial y} = \frac{\partial}{\partial y}(y) = 1,$$

$$\frac{\partial Q}{\partial x} = \frac{\partial}{\partial x}(2x) = 2,$$

(2.4.37)

therefore they differ. By theorem 2.2, the integral is **path dependent.** Even if the integral had the same value on some paths with the same ends, this would just happen by chance!

In this case, we say that **the force F is not conservative.**

111

*Example.* Compute the circulation of the vector

$$\mathbf{V}\left( \underbrace{x^2 y}_{P}, \underbrace{x^3}_{Q} \right)$$ along the closed contour $K$, formed by the

intersection of the parabolas $y = x^2$, $x = y^2$, which runs counterclockwise.

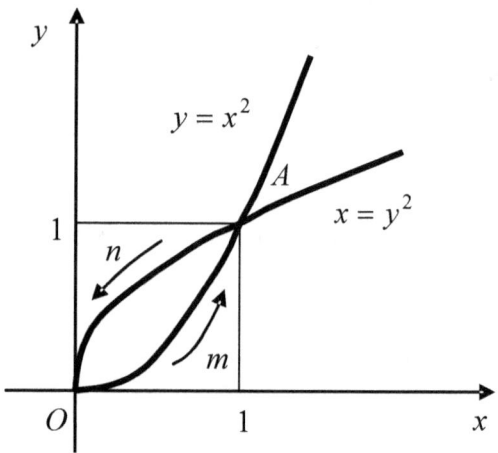

*Figure 2. 14. The intersection of the parabolas $y = x^2$, $x = y^2$*

**Solution.** The circulation is computed by the formula:

$$c = \oint_K P(x, y)\, dx + Q(x, y)\, dy = \oint_K x^2 y\, dx + x^3\, dy =$$
$$= \int_{\overset{\frown}{OmA}} x^2 y\, dx + x^3\, dy + \int_{\overset{\frown}{AnO}} x^2 y\, dx + x^3\, dy. \qquad (2.4.38)$$

The arc $\overset{\frown}{OmA}$ is expressed as follows

$$\overset{\frown}{OmA} : \begin{cases} x = x \\ y = x^2 \end{cases} \quad x \in [0,1]. \qquad (2.4.39)$$

It follows that $dy = 2x\, dx$ and

112

$$\int_{\overset{\frown}{OmA}} x^2\, y\, \mathrm{d}x + x^3\, \mathrm{d}y = \int_0^1 \left(x^2 \cdot x^2 + x^3 \cdot 2\, x\right) \mathrm{d}x =$$

$$= \int_0^1 3\, x^4\, \mathrm{d}x = \frac{3\, x^5}{5}\Big|_0^1 = \frac{3}{5}. \tag{2.4.40}$$

The equations of the arc $\overset{\frown}{OnA}$ are

$$\overset{\frown}{OnA} : \begin{cases} y = y \\ x = y^2 \end{cases} \quad y \in [0,1]. \tag{2.4.41}$$

It follows that $\mathrm{d}x = 2y\,\mathrm{d}y$ and

$$\int_{\overset{\frown}{AnO}} x^2\, y\, \mathrm{d}x + x^3\, \mathrm{d}y = - \int_{\overset{\frown}{OnA}} x^2\, y\, \mathrm{d}x + x^3\, \mathrm{d}y =$$

$$= -\int_0^1 \left(y^4 \cdot y \cdot 2\, y + y^6\right) \mathrm{d}y = -\int_0^1 3\, y^6\, \mathrm{d}y = -\frac{3\, y^7}{7}\Big|_0^1 = -\frac{3}{7}. \tag{2.4.42}$$

Finally, the circulation is given by

$$c = \frac{3}{5} - \frac{3}{7} = \frac{6}{35}. \tag{2.4.43}$$

### 2.4.5. VECTOR FORM

Let **F** be a force of components $P$, $Q$ and $\overset{\frown}{AB}$ an arc included in $D$.

***The mechanical work*** done by $\mathbf{F}(P,Q)$ is

$$\mathcal{L}_{\overset{\frown}{AB}}(\mathbf{F}) = \int_{\overset{\frown}{AB}} P(x,y)\mathrm{d}x + Q(x,y)\mathrm{d}y. \tag{2.4.44}$$

But $\mathbf{F} = P\mathbf{i} + Q\mathbf{j}$. If $\mathbf{r} = x\mathbf{i} + y\mathbf{j}$ is the position vector in the plane, then $d\mathbf{r} = \mathbf{i}dx + \mathbf{j}dy$, therefore $\mathbf{F} \cdot d\mathbf{r} = Pdx + Qdy$. It follows that

$$\boxed{\mathcal{L}_{\overset{\frown}{AB}}(\mathbf{F}) = \int_{\overset{\frown}{AB}} \mathbf{F} \cdot d\mathbf{r}}. \qquad (2.4.45)$$

Let $\mathbf{V}(P,Q)$ be a plane field. Then, as in the case of the mechanical work, we can show that the circulation of $\mathbf{V}$ along the arc $\overset{\frown}{AB}$ can be put in vector form as follows

$$\boxed{c_{\overset{\frown}{AB}}(\mathbf{V}) = \int_{\overset{\frown}{AB}} \mathbf{V} \cdot d\mathbf{r}}. \qquad (2.4.46)$$

# 2.5. SECOND KIND CURVILINEAR INTEGRALS IN SPACE

Let $P, Q, R \in C^0(\overset{\frown}{AB})$, $\overset{\frown}{AB} \subset \mathfrak{R}^3$. Then, the curvilinear integral of the second kind

$$I = \int_{\overset{\frown}{AB}} P(x, y, z)dx + Q(x, y, z)dy + R(x, y, z)dz, \qquad (2.5.1)$$

is defined by generalizing the one from the two-dimensional case (using again the Riemann sums).

## 2.5.1. THE CALCULATION OF THE SECOND KIND CURVILINEAR INTEGRAL IN SPACE

Consider the arc

$$AB : \begin{cases} x = x(t) \\ y = y(t), \quad t \in [a,b] \subset \Re. \\ z = z(t) \end{cases} \quad (2.5.2)$$

We assume that this arc is smooth, which means that $(x, y, z) \in C^1 ([a,b])$.

Then one can show that the second kind curvilinear integral is computed by the formula

$$I = \int_a^b [P(x(t), y(t), z(t))x'(t) +$$
$$+ Q(x(t), y(t), z(t))y'(t) + \quad (2.5.3)$$
$$+ R(x(t), y(t), z(t))z'(t)]dt.$$

The right hand side of the equation is a Riemann integral, computed on the segment $[a,b]$.

## 2.5.2. PROPERTIES OF THE SECOND KIND CURVILINEAR INTEGRAL IN SPACE

They are the same as in the plane case.

### 2.5.3. VECTOR FORM

Let $\mathbf{V}(P,Q,R)$ be a vector field; we assume that $P, Q, R \in C^1(\Omega)$ on the tridimensional domain $\Omega$. If $\mathbf{r} = x\mathbf{i} + y\mathbf{j} + z\mathbf{k}$ is the position vector, then:

$$d\mathbf{r} = \mathbf{i}\,dx + \mathbf{j}\,dy + \mathbf{k}\,dz, \quad (2.5.4)$$

and

$$\mathbf{V} \cdot d\mathbf{r} = P\,dx + Q\,dy + R\,dz. \qquad (2.5.5)$$

Therefore

$$I = \int_{\overset{\frown}{AB}} \mathbf{V} \cdot d\mathbf{r}, \quad \overset{\frown}{AB} \subset \Omega. \qquad (2.5.6)$$

## 2.5.4. PHYSICAL INTERPRETATION

**1.** Let us consider the force $\mathbf{F}(P,Q,R)$. We assume that its components $P,Q,R$ are continuous on the skew arc $\overset{\frown}{AB}$. Then

$$\mathcal{L}_{\overset{\frown}{AB}}(\mathbf{F}) = \int_{\overset{\frown}{AB}} \mathbf{F} \cdot d\mathbf{r} \qquad (2.5.7)$$

is the **mechanical work** done by the force $\mathbf{F}$ along the arc $\overset{\frown}{AB}$.

**2.** Let $\mathbf{V}(P,Q,R)$ be a vector field. Suppose $P,Q,R \in C^0(\overset{\frown}{AB})$, where $\overset{\frown}{AB}$ is a skew arc. Then

$$c_{\overset{\frown}{AB}} = \int_{\overset{\frown}{AB}} \mathbf{V} \cdot d\mathbf{r} \qquad (2.5.8)$$

represents the **circulation** of the vector $\mathbf{V}$ along $\overset{\frown}{AB}$.

If $\Gamma \subset \mathfrak{R}^3$ is a closed contour, then we use the following notation

$$\boxed{I = \oint_{\Gamma} \mathbf{V} \cdot d\mathbf{r}}. \qquad (2.5.9)$$

In what follows, we generalize the theorem 2.2 (path independence):

**Theorem 2.3.** *Let* $\mathbf{V}(P,Q,R) \in [C^1(\Omega)]^3$, $\Omega \subset \mathcal{R}^3$. *Then*

$$I = \int_{\overset{\frown}{AB}} \mathbf{V} \cdot d\mathbf{r} \text{ is \textbf{path independent} if and only if } \operatorname{rot}\mathbf{V} = 0 \text{ in } \Omega.$$

* **Proof.** As in the theorem 2.2,

$$\operatorname{rot}\mathbf{V} = \begin{vmatrix} \mathbf{i} & \mathbf{j} & \mathbf{k} \\ \dfrac{\partial}{\partial x} & \dfrac{\partial}{\partial y} & \dfrac{\partial}{\partial z} \\ P & Q & R \end{vmatrix} = \quad (2.5.10)$$

$$= \mathbf{i}\left(\frac{\partial R}{\partial y} - \frac{\partial Q}{\partial z}\right) - \mathbf{j}\left(\frac{\partial R}{\partial x} - \frac{\partial P}{\partial z}\right) + \mathbf{k}\left(\frac{\partial Q}{\partial x} - \frac{\partial P}{\partial y}\right).$$

Therefore $\operatorname{rot}\mathbf{V} = 0$ if and only if the equalities

$$\frac{\partial R}{\partial y} = \frac{\partial Q}{\partial z}, \quad \frac{\partial R}{\partial x} = \frac{\partial P}{\partial z}, \quad \frac{\partial Q}{\partial x} = \frac{\partial P}{\partial y} \qquad (2.5.11)$$

occur simultaneously.

While not obviously, the last equality also appears in the theorem 2.2. In fact, in the particular case of a plane vector, we have $\mathbf{V}(P(x,y),Q(x,y),0)$. Therefore $\operatorname{rot}\mathbf{V} = \mathbf{k}\left(\dfrac{\partial Q}{\partial x} - \dfrac{\partial P}{\partial y}\right)$.

Hence, in the plane case, the relations (2.5.11) are reduced to the hypothesis of the theorem 2.2.

*Example.* Compute the mechanical work done by the force

$\mathbf{F}(xy,\ yz,\ zx)$ along the arc $\overset{\frown}{AB}$, of equations

$$\begin{cases} x = \cos t \\ y = \sin t, \ t \in \left[0, \dfrac{\pi}{2}\right]. \\ z = 1 \end{cases}$$

**Do we get the same value for** $\mathcal{L}_{\overset{\frown}{AB}}(\mathbf{F})$ **if we replace the**

**path** $\overset{\frown}{AB}$ **with another path having the same ends** $A, B$ **?**

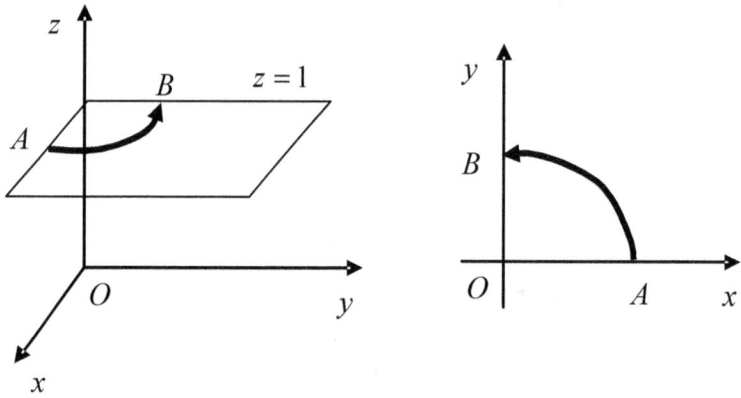

*Figure 2. 15. The representation of the arc in the plane* $z = 1$

**Solution.** As we notice from the figure 2.15, the arc $\overset{\frown}{AB}$
is situated in the plane $z = 1$. We have

$$\mathcal{L}_{\overset{\frown}{AB}}(\mathbf{F}) = \int_{\overset{\frown}{AB}} \mathbf{F} \cdot d\mathbf{r} = \int_a^b xy\,dx + y\underset{=1}{\underline{z}}\,dy + \underset{=1}{\underline{z}}\,x\underset{=0}{\underline{dz}} =$$

$$= \int_0^{\frac{\pi}{2}} \left[\cos t \sin t \cdot (-\sin t) + \sin t \cdot \cos t\right] dt = \qquad (2.5.12)$$

$$= \frac{\sin^3 t}{3}\Bigg|_{t=0}^{t=\frac{\pi}{2}} + \frac{\sin^2 t}{2}\Bigg|_{t=0}^{t=\frac{\pi}{2}},$$

such that, finally,

$$\boxed{\mathcal{L}_{\underset{AB}{\frown}}(\mathbf{F}) = \frac{1}{6}}.$$

(2.5.13)

In order to see if the value of $\mathcal{L}_{\underset{AB}{\frown}}(\mathbf{F})$ is preserved by changing the path, still keeping its ends, we compute

$$\text{rot}\,\mathbf{V} = \begin{vmatrix} \mathbf{i} & \mathbf{j} & \mathbf{k} \\ \dfrac{\partial}{\partial x} & \dfrac{\partial}{\partial y} & \dfrac{\partial}{\partial z} \\ xy & yz & zx \end{vmatrix} = \mathbf{i}(-y) - \mathbf{j}\,z + \mathbf{k}(-x) \neq \mathbf{0}.$$

(2.5.14)

Thus, **the value of** $\mathcal{L}_{\underset{AB}{\frown}}(\mathbf{F})$ **is not preserved**, therefore the force **F** is **not conservative**.

## * 2.6. THE LINK BETWEEN THE CURVILINEAR INTEGRALS OF THE FIRST AND SECOND KIND

**In plane**: Consider the arc $\overset{\frown}{AB} : \begin{cases} x = x(t) \\ y = y(t) \end{cases}, \; t \in [a,b],$

situated in the plane $xOy$. Take on this arc the points $M(x,y)$, $M'(x + \Delta x, y + \Delta y)$, corresponding to $t, t + \Delta t$ respectively ( figure 2.16).

The direction cosines of $\overline{MM'}$ are

$$\frac{\Delta x}{\sqrt{\Delta x^2 + \Delta y^2}}, \; \frac{\Delta y}{\sqrt{\Delta x^2 + \Delta y^2}},$$

(2.5.15)

or, if the arc is expressed in terms of a parameter $t$,

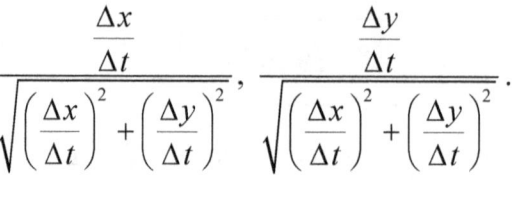

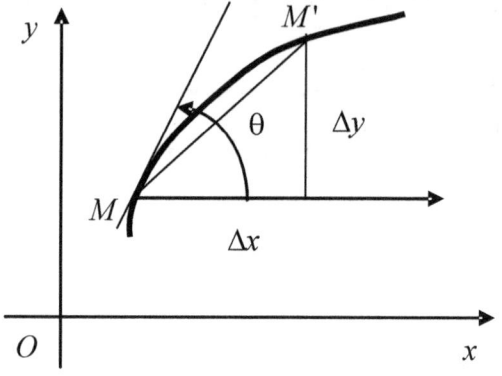

*Figure 2. 16. The position of the elements for the connection between the two curvilinear integrals in plane*

Taking the limit for $M' \to M$ in (2.5.15), we get respectively $\dfrac{\mathrm{d}x}{\mathrm{d}s}, \dfrac{\mathrm{d}y}{\mathrm{d}s}$, where $\mathrm{d}s = \sqrt{\mathrm{d}x^2 + \mathrm{d}y^2}$ is the element of arc length. We can also write

$$\alpha = \frac{\mathrm{d}x}{\mathrm{d}s} = \cos\theta, \quad \beta = \frac{\mathrm{d}y}{\mathrm{d}s} = \sin\theta, \tag{2.5.16}$$

where $\theta$ is the angle made by the tangent in $M$ to $\overset{\frown}{AB}$ with the positive direction of the $Ox$ axis.

Therefore, if $\boldsymbol{\tau} = (\alpha, \beta)$ and $\mathbf{F}(P,Q)$ is a vector of class $C^0(D)$, $D$ being a plane domain including $\overset{\frown}{AB}$, then

$$\int_{\overset{\frown}{AB}} P(x,y)dx + Q(x,y)dy =$$

$$= \int_{\overset{\frown}{AB}} \left[ P(x,y)\alpha + Q(x,y)\beta \right] ds. \qquad (2.5.17)$$

The left hand member is the integral of the second kind and the right hand one is the integral of the first kind.

In vector form, we have

$$\boxed{\int_{\overset{\frown}{AB}} \mathbf{F} \cdot d\mathbf{r} = \int_{\overset{\frown}{AB}} \mathbf{F} \cdot \boldsymbol{\tau} ds.} \qquad (2.5.18)$$

***In space***: Using a similar proof, the direction cosines of the plane tangent to the skew arc $\overset{\frown}{AB}$ at the point $M$ are obtained from

$$\frac{\Delta x}{\sqrt{\Delta x^2 + \Delta y^2 + \Delta z^2}}, \quad \frac{\Delta y}{\sqrt{\Delta x^2 + \Delta y^2 + \Delta z^2}},$$

$$\frac{\Delta z}{\sqrt{\Delta x^2 + \Delta y^2 + \Delta z^2}} \qquad (2.5.19)$$

( figure 2.17).

Taking limits, we obtain the following expressions

$$\alpha = \frac{dx}{ds}, \quad \beta = \frac{dy}{ds}, \quad \gamma = \frac{dz}{ds}, \qquad (2.5.20)$$

for the direction cosines of the plane tangent to $\overset{\frown}{AB}$ at $M$.

If $P, Q, R \in C^0(\Omega)$, where $\Omega$ is a tridimensional domain containing the arc $\overset{\frown}{AB}$, it follows that

$$\int\limits_{\overset{\frown}{AB}} P(x,y,z)dx + Q(x,y,z)dy + R(x,y,z)dz =$$

$$= \int\limits_{\overset{\frown}{AB}} \left( P(x,y,z)\frac{dx}{ds} + Q(x,y,z)\frac{dy}{ds} + R(x,y,z)\frac{dz}{ds} \right)ds = \quad (2.5.21)$$

$$= \int\limits_{\overset{\frown}{AB}} [P(x,y,z)\alpha + Q(x,y,z)\beta + R(x,y,z)\gamma]ds,$$

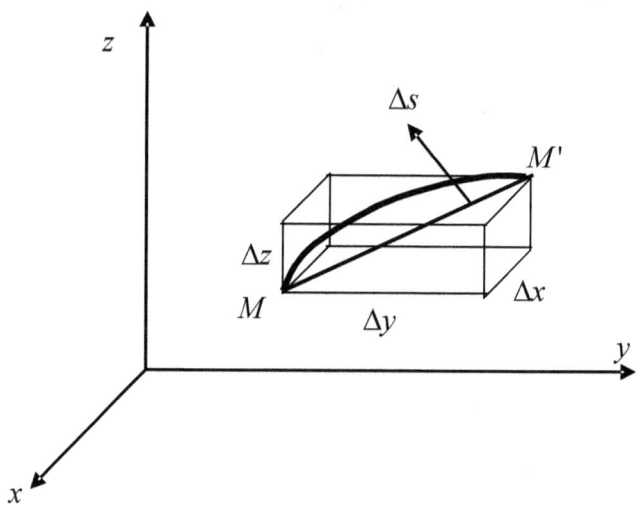

*Figure 2. 17. The position of the elements for the connection between the two curvilinear integrals on skew curves*

the left hand being a second kind and the last one being a first kind integral.

Using the vector form, if $\mathbf{F} = \mathbf{F}(P,Q,R)$ and $\mathbf{r} = x\mathbf{i} + y\mathbf{j} + z\mathbf{k}$ is the position vector, then, denoting by $\boldsymbol{\tau} = (\alpha, \beta, \gamma)$, we get

$$\int\limits_{\overset{\frown}{AB}} \mathbf{F} \cdot d\mathbf{r} = \int\limits_{\overset{\frown}{AB}} \mathbf{F} \cdot \boldsymbol{\tau} ds. \quad (2.5.22)$$

The two expressions (2.5.18) and (2.5.22) formally coincide; this was predictable.

## EXERCISES AND PROBLEMS

### *FIRST KIND CURVILINEAR INTEGRALS*

1. Compute the following curvilinear integrals of the first kind:

a) $I = \oint_C \left( x^2 + y^2 \right)^3 ds, \ C : x^2 + y^2 = a^2$   A: $I = 2\pi a^7$

b) $I = \oint_C \left( x^{\frac{4}{3}} + y^{\frac{4}{3}} \right) ds, \ C : x^{\frac{2}{3}} + y^{\frac{2}{3}} = a^{\frac{2}{3}}$   A: $I = 4a^{\frac{7}{3}}$
Hint: See ex.4

c) $I = \oint_C \sqrt{x^2 + y^2} \ ds, \ C : x^2 + y^2 - ax = 0$

A: $I = 2a^2$

Hint:The parametric equations are:

$$\begin{cases} x = \dfrac{a}{2} + \dfrac{a}{2}\cos t \\ y = \dfrac{a}{2}\sin t \end{cases}, 0 \le t \le 2\pi$$

d) $I = \int_{\overset{\frown}{AB}} x^2 y^2 ds, \ \overset{\frown}{AB} : \begin{cases} x = a\cos^3 t \\ y = a\sin^3 t \end{cases}, 0 \le t \le 2\pi$   A: $I = 0$

e) $I = \int_{\overset{\frown}{AB}} xyzds, \ \overset{\frown}{AB} : \begin{cases} x = t \\ y = t^2 \\ z = \dfrac{2}{3}t^3 \end{cases}, 0 \le t \le 1$   A: $I = \dfrac{46}{189}$

f)
$$I_1 = \int_{\overset{\frown}{AB}} \left(x^2 + y^2 + z^2\right) ds,$$

$$I_2 = \int_{\overset{\frown}{AB}} \frac{1}{x^2 + y^2 + z^2} ds,$$

$$AB : \begin{cases} x = a\cos t \\ y = a\sin t \quad ,0 \le t \le 2\pi \\ z = bt, b > 0 \end{cases}$$

A:
$$I_1 = \frac{2\pi}{3} \left(3a^2 + 4\pi^2 b^2\right)\sqrt{a^2 + b^2}$$

$$I_2 = \frac{\sqrt{a^2 + b^2}}{ab} \operatorname{arctg} \frac{2\pi b}{a}$$

g) $I = \int_{\overset{\frown}{AB}} \left(x^2 + y^2\right)\ln z \, ds,$ $AB : \begin{cases} x = e^t \cos t \\ y = e^t \sin t, 0 \le t \le 1 \\ z = e^t \end{cases}$

A: $I = \dfrac{\sqrt{3}\left(2e^3 + 1\right)}{9}$

h) $I = \int_{\overset{\frown}{AB}} z\left(x^2 + y^2\right) ds,$ $AB : \begin{cases} x = t\cos t \\ y = t\sin t, 0 \le t \le 1 \\ z = t \end{cases}$

A: $I = \dfrac{8\sqrt{2} - 3\sqrt{3}}{15}$

2. Find the length of the circle of equation $x^2 + y^2 = a^2$.

A: $l_c = 2\pi a$

3. Compute the mass of the wire of a triangle form with the edges $O(0,0), A(1,0), B(0,1)$, if its density is $\rho(x, y) = x + y$, $\rho > 0$.

A: $m = 1 + \sqrt{2}$

4. Compute the mass of the arc $\overset{\frown}{AB}$ of density

$$\rho(x, y) = \sqrt{2y}, \rho > 0,$$

where $\overset{\frown}{AB}$ is the cycloid loop $\begin{cases} x = a(t - \sin t) \\ y = a(1 - \cos t) \end{cases}$, $0 \le t \le 2\pi$.

$$A: m_{\overset{\frown}{AB}} = 4\pi a\sqrt{a}$$

5. Find the length of the astroid quarter $x^{\frac{2}{3}} + y^{\frac{2}{3}} = a^{\frac{2}{3}}$, placed in the first quadrant ($x \ge 0$, $y \ge 0$).

$$A: l_{\overset{\frown}{AB}} = \frac{3a}{2}$$

*Hint*: The parametric equations of the astroid are:

$$\begin{cases} x = a\cos^3 t \\ y = a\sin^3 t \end{cases}, t \in \left[0, \frac{\pi}{2}\right].$$

6. Find the length of the arc $\overset{\frown}{AB}$, expressed by the parametric equations $\overset{\frown}{AB}: \begin{cases} x = a(1 - \cos t) \\ y = a(t - \sin t) \end{cases}, t \in \left[0, \frac{\pi}{2}\right], a > 0$.

$$A: l_{\overset{\frown}{AB}} = 4a\left(1 - \frac{\sqrt{2}}{2}\right)$$

7. Compute the mass of the arc $\overset{\frown}{AB}$, of density $\rho(x, y) = xy$, $\rho > 0$, where $\overset{\frown}{AB}$ is the quarter of the ellipse $\frac{x^2}{a^2} + \frac{y^2}{b^2} = 1$, placed in the first quadrant.

$$A: m_{\overset{\frown}{AB}} = \frac{ab(a^2 + ab + b^2)}{3(a + b)}$$

*Hint:* $\overset{\frown}{AB}: \begin{cases} x = a\cos t \\ y = b\sin t \end{cases}, \quad t \in \left[0, \dfrac{\pi}{2}\right]$

8. Find the length of the conical spiral, expressed by the

parametric equations $\begin{cases} x = e^t \cos t \\ y = e^t \sin t, \quad t \in [0, 2\pi]. \\ z = e^t \end{cases}$

$$\text{A:} l = \sqrt{3}\left(e^{2\pi} - 1\right)$$

## CURVILINEAR INTEGRALS OF THE SECOND KIND

1. Compute the following curvilinear integrals of the second kind:

a) $I = \displaystyle\int_{\overset{\frown}{AB}} \dfrac{y\,dx - x\,dy}{x^2}$, along the path which joins the points

$A(2,1)$ and $B(1,2)$, without crossing the $Oy$ axis.

$$\text{A:} I = -\dfrac{3}{2}$$

*Hint:* Verify the path independence.

b) $I = \displaystyle\int_{\Gamma} \dfrac{y\,dx + x\,dy}{x^2 + y^2}$, where $\Gamma = \overset{\frown}{AB} \cup \overline{BC}$, $\overset{\frown}{AB}$ is the

demi-circle centered at the origin, and $\overline{BC}$ is the segment joining the points $B$ and $C$ and the points $A$, $B$, $C$ are of coordinates $A(0,2)$, $B(2,0)$, $C(2,-2)$.

$$\text{A:} I = -\dfrac{\pi}{4}$$

c) $I = \int\limits_{\overset{\frown}{AB}} \dfrac{x^2\,dy - y^2\,dx}{x^{\frac{5}{3}} + y^{\frac{5}{3}}}$, where $\overset{\frown}{AB}$ is the arc of the

astroid $x^{\frac{2}{3}} + y^{\frac{2}{3}} = a^{\frac{2}{3}}$, $A(a,0)$, and $B(0,b)$.

$$A:\ I = \dfrac{3a^{\frac{4}{3}}\pi}{16}$$

*Hint*: The parametric equations of the astroid are:

$$\begin{cases} x = a\cos^3 t \\ y = a\sin^3 t \end{cases},\ t \in \left[0, \dfrac{\pi}{2}\right].$$

d) $I = \int\limits_{\overset{\frown}{OA}} (2a - y)\,dx + x\,dy$, $\overset{\frown}{OA}$ being the cycloid loop of

parametric equations $\begin{cases} x = a(t - \sin t) \\ y = a(1 - \cos t) \end{cases}$, $0 \le t \le 2\pi$.

$$A:I = -2\pi a^2$$

e) $I = \int\limits_{\overset{\frown}{AB}} (x^2 - 2xy)\,dx + (y^2 - 2xy)\,dy$, $\overset{\frown}{AB}$ being the arc

of parabola $y = x^2$, whose sense is given by the growth of

$x \in [-1,1]$.

$$A:I = -\dfrac{14}{15}$$

f) $I = \int\limits_{C} \dfrac{dx + dy}{|x| + |y|}$, $C$ being the contour of the square of

edges $A(1,0)$, $B(0,1)$, $C(-1,0)$, $D(0,-1)$.

$$A: I = 0$$

g) $I = \int\limits_{\overset{\frown}{AB}} y\,dx - z\,dy + x\,dz$, where $\overset{\frown}{AB}$ is the arc of

helicoid defined by the parametric equations

$$\begin{cases} x = a\cos t \\ y = a\sin t \quad, 0 \le t \le 2\pi. \\ z = bt, b > 0 \end{cases}$$

$$A: I = -\pi a^2$$

2. Compute the mechanical work done by the force of components $\overrightarrow{F}\left(-(x+y), x\right)$ along the segment $\overline{AB}$ joining the points $A(1,0)$ and $B(0,1)$.

$$A: \mathcal{L}_{\overline{AB}}(\mathbf{F}) = \frac{3}{2}$$

3. Compute the mechanical work done by the force of components $\overrightarrow{F}\left(x^2 + y^2, xy\right)$ along the parabola $y = x^2$ which joins the points $O(0,0)$ and $A(1,1)$.

$$A: \mathcal{L}_{\overline{OA}}(\mathbf{F}) = \frac{14}{15}$$

4. Compute the mechanical work done by the force of components $\mathbf{F}\left(x + 2y, 2x + y\right)$ along a path which joins the points $O(0,0)$ and $A(2,3)$.

$$A: \mathcal{L}_{\overline{OA}}(\mathbf{F}) = \frac{37}{2}$$

*Hint*: Verify the path independence.

5. Find the parameter $k$ such that the curvilinear integral

$$I = \int_{\overset{\frown}{AB}} \frac{1+ky^2}{(1+xy)^2}dx + \frac{1+kx^2}{(1+xy)^2}dy$$

be path independent.

A: $k = -1$

6. Compute the mechanical work done by the force of components $\mathbf{F}(x^2+y^2, x^2-y^2)$ along the curve $y = 1-|1-x|$, oriented towards the argument growth $x \in [0,2]$.

A: $\mathcal{L}_{\overset{\frown}{AB}}(\mathbf{F}) = \dfrac{4}{3}$

7. Compute the mechanical work done by the force of components $\mathbf{F}(y+1, x^2)$ along the parabola $y = x^2 - 1$, $x \in [-1,1]$ whose sense is given by the growth of $x$.

A: $\mathcal{L}_{\overset{\frown}{AB}}(\mathbf{F}) = \dfrac{2}{3}$

8. Compute the mechanical work done by the force of components $\mathbf{F}(y,x)$ along the path which joins the points $A(-1,2)$ and $B(2,3)$.

A: $\mathcal{L}_{\overset{\frown}{AB}}(\mathbf{F}) = 8$

*Hint*: Verify the path independence.

9. Compute the circulation of the vector $\overline{V}(x+y, x-y)$ along the ellipse $\dfrac{x^2}{a^2} + \dfrac{y^2}{b^2} = 1$, run in trigonometric sense.

A: $c = 0$

10. Compute the mechanical work done by the force of the components $\mathbf{F}\left(\dfrac{y}{1+xy}, \dfrac{x}{1+xy}\right)$ along the path which joins the points $A\left(\dfrac{1}{3}, -2\right)$ and $B(3,0)$, and does not cross the hyperbola $xy = -1$.

A: $\underset{AB}{\mathcal{L}}(\mathbf{F}) = \ln 3$

*Hint*: Verify the path independence and choose the simplest path which respects the condition of the problem. This condition is imposed by the denominator of the expressions of the components of $\mathbf{F}$.

11. Compute the mechanical work done by the force of components $\mathbf{F}(-xy, y(x+1))$ along the part of the ellipse from the superior semi-plane, $\dfrac{x^2}{a^2} + \dfrac{y^2}{b^2} = 1, y \geq 0$.

A: $\underset{AB}{\mathcal{L}}(\mathbf{F}) = \dfrac{2ab^2}{3}$

12. Compute the mechanical work done by the force of components $\mathbf{F}(x, y, z)$ along the arc $\overset{\frown}{AB}$ defined by the parametric equations: $\begin{cases} x = a\cos t \\ y = a\sin t, 0 \leq t \leq 2\pi. \\ z = bt \end{cases}$

A: $\underset{AB}{\mathcal{L}}(\mathbf{F}) = 2b^2\pi^2$

13. Compute the mechanical work done by the force of components $\mathbf{F}\left(y^2 - z^2, 2yz, -x^2\right)$ along the arc $\overset{\frown}{AB}$ of equations

$$\begin{cases} x = t \\ y = t^2, \\ z = t^3 \end{cases} \quad 0 \leq t \leq 1, \text{ whose sense is given by the growth of } t.$$

A: $\underset{\overset{\frown}{AB}}{\mathcal{L}}(\mathbf{F}) = \dfrac{1}{35}$

# Chapter 3

## THE DOUBLE INTEGRAL

### 3.1. THE AREA OF A PLANE DOMAIN

We start from the area of a rectangle, as a basis.

**Definition 3.1.** The set $D \subset \Re^2$ is called ***elementary*** if $D = \bigcup_{i=1}^{n} D_i$, where

1. $D_i$ are rectangles with sides parallel to the axes of coordinates;

2. $D_i, D_j, \, i \neq j$, have, at most, one side in common.

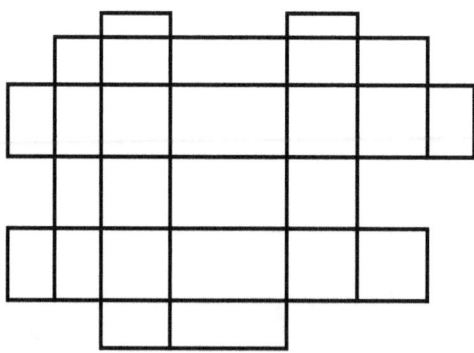

*Figure 3. 1. Example of elementary set*

We denote the area of $D$ by $|D|$; we know how to compute it, obviously, we sum the areas of the rectangles which compose it.

**Definition 3.2.** We say that the bounded set $A \subset \mathfrak{R}^2$ *has an area* if one can find two sequences $\{E_n\}_{n \in \mathfrak{N}}$, $\{F_n\}_{n \in \mathfrak{N}}$ of elementary sets such that:

1. $F_n \subset A \subset F_n$, $\forall n \in \mathfrak{N}$;

2. The sequences of positive numbers $\{|E_n|\}_{n \in \mathfrak{N}}$, $\{|F_n|\}_{n \in \mathfrak{N}}$ converge to the same limit

$$l = \lim_{n \to \infty} |E_n| = \lim_{n \to \infty} |F_n|. \qquad (3.1.1)$$

Then, *the area of the set* $A$ is $\boxed{|A| = l}$.

## 3.2. THE DEFINITION OF THE DOUBLE INTEGRAL

Let $D \subset \mathfrak{R}^2$ be a closed bounded plane domain, of area $|D|$. We consider a partition $\Delta$ (figure 3.2), formed by $n$ subdomains $D_i$, $i = \overline{1,n}$, of areas $|D_i| = \Delta\sigma_i$, with the following properties:

1. $D = \bigcup_{i=1}^{n} D_i$;

2. $D_i$, $D_j$, $i \neq j$ have, at most, boundary points in common.

**Definition 3.3.** We define *the diameter* of a set $A$ as the maximum distance between two points of the boundary of $A$ (denoted by *fr A*):

$$\text{diam}\, A = \max \left\{ |\overline{PQ}|, \ P, Q \in \text{fr } A \right\}. \qquad (3.2.1)$$

*Examples of diameters:*

**a)** The ***diameter of a circle*** $C$ is $\operatorname{diam} C = \left| \overline{AB} \right|$, i.e., the length of the diameter of the circle.

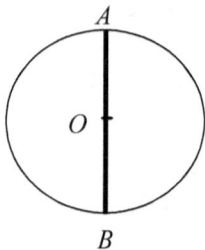

*Figure 3. 2 The diameter of a circle*

**b)** The ***diameter of a rectangle*** $\Omega$ is $\operatorname{diam} \Omega = \left| \overline{AC} \right|$, i.e., the length of its diagonal.

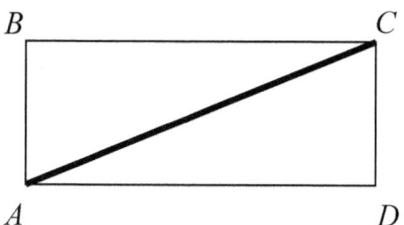

*Figure 3. 3 The diameter of a rectangle*

**c)** The ***diameter of an ellipse*** $E$ is $\operatorname{diam} E = 2a$, where $a$ is the length of the semi-major axis of the ellipse.

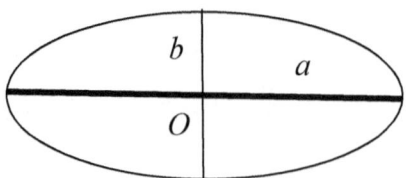

*Figure 3. 4 The diameter of an ellipse*

Now, let $d_1, d_2, \ldots, d_n$ be the diameters of the sets $D_1, D_2, \ldots, D_n$ of the partition $\Delta$ of $D$.

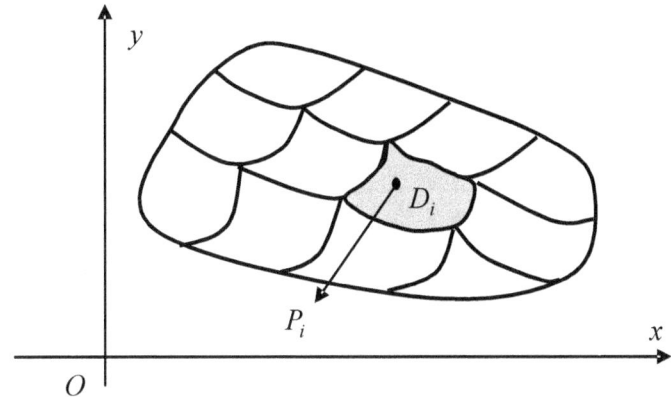

*Figure 3.5. Partition on the closed bounded domain D*

Therefore $\operatorname{diam} D_i = d_i$, $i = \overline{1, n}$. The **norm** (mesh size) of the partition $\Delta$ is

$$v(\Delta) = \max \left\{ d_i,\ i = \overline{1, n} \right\}. \tag{3.2.2}$$

Let $f : D \to \Re$. In each $D_i$ we choose an arbitrary point $P_i(\alpha_i, \beta_i)$. Now, we can set up the Riemann sum

$$\sigma_\Delta(f, P_i) = \sum_{i=1}^{n} f(\alpha_i, \beta_i) \cdot \Delta\sigma_i, \tag{3.2.3}$$

corresponding to the partition $\Delta$ and to the choice $\{P_i\}_{i=\overline{1,n}}$ of the intermediate points.

**Definition 3.4.** If there exists a real number $I \in \Re$ with the property that for any $\varepsilon > 0$ one can find $\eta(\varepsilon)$ such that the inequality $|\sigma_\Delta(f, P_i) - I| < \varepsilon$ be valid for any partition $\Delta$ of norm $v(\Delta) < \eta$

and for any choice of the intermediate points $P_i$, then $f$ is called *integrable on* $D$ and

$$I = \iint_D f(x, y) \, d\sigma \qquad (3.2.4)$$

is *the double integral of $f$ on $D$.*

In other words, $I = \lim_{\substack{v(\Delta) \to 0, \\ \forall P_i}} \sigma_\Delta(f, P_i)$.

*Remark.* The definition 3.4 has the same logical pattern as the definition 1.2. This logical pattern is the same for any Riemann integral, the only difference occuring in the partitions and in their norms. We shall find this pattern in all the other types of integrals defined in what follows.

### 3.2.1. THE DARBOUX SUMS

If $f$ is bounded on $D$, we can also use the following sums in the case of the double integral:

♣ **The upper Darboux sum:**

$$S_\Delta(f) = \sum_{i=1}^{n} M_i \Delta\sigma_i, \quad M_i = \sup_{(x, y) \in D_i} f(x, y);$$

♣ **The lower Darboux sum:**

$$s_\Delta(f) = \sum_{i=1}^{n} m_i \Delta\sigma_i, \quad m_i = \inf_{(x, y) \in D_i} f(x, y).$$

The following inequality is valid

$$s_\Delta(f) < \sigma_\Delta(f, P_i) < S_\Delta(f). \qquad (3.2.5)$$

Using the same methods as those from the Riemann integral on a real interval, we can prove *the Darboux criterion:*

**Theorem 3.1.** *Let f be a function defined and bounded on the closed bounded domain* $D \subset \mathfrak{R}^2$. *Then f is integrable on D if and only if for any* $\varepsilon > 0$ *there exists a number* $\eta(\varepsilon)$ *such that* $S_\Delta(f) - s_\Delta(f) < \varepsilon$ *for any partition of norm* $v(\Delta) < \eta$.

A self-understanding corollary of this theorem is:

**Theorem 3.2.** *Let f be a function continuous on the closed and bounded domain* $D \subset \mathfrak{R}^2$ *(i.e.,* $f \in C^0(D)$*). Then f is integrable on D.*

## 3.2.2. PROPERTIES OF THE DOUBLE INTEGRAL

We shall present several properties of the double integral which can be easily deduced from the corresponding Riemann sums.

Let $D \subset \mathfrak{R}^2$ be a compact domain; as $\mathfrak{R}^2$ is finite-dimensional, the property of compactness is equivalent to the property of being closed and bounded.

**1.** If $|D| = 0$, then

$$\iint_D f(x, y) \, d\sigma = 0, \tag{3.2.6}$$

for any $f$ (integrable on $D$).

**Proof.** For any partition $\Delta$ of $D$, the corresponding Riemann sums are, in this case,

$$\sigma_\Delta(f, P_i) = \sum_{i=1}^{n} f(\xi_i, \eta_i) \Delta\sigma_i = 0, \tag{3.2.7}$$

for any choice of the intermediate points $P_i$, whence it follows that $\lim_{v(\Delta) \to 0} \sigma_\Delta(f, P_i)$ exists and is zero. ◻

**2.** For $f = 1$ on a compact domain $D \subset \Re^2$, we have

$$\iint_D d\sigma = |D|. \tag{3.2.8}$$

**Proof.** For any partition $\Delta$ and any choice of the intermediate points we have

$$\sigma_\Delta(f, P_i) = \sum_{i=1}^n 1 \cdot \Delta\sigma_i = |D|. \; \blacksquare \tag{3.2.9}$$

**3. PROPERTIES OF MONOTONICITY**

*a)* If $f \geq 0$ is integrable on $D$, then

$$\iint_D f(x, y) d\sigma \geq 0. \tag{3.2.10}$$

**Proof.** The associated Riemann sums are, all of them, positive.

*b)* If $f, g$ are integrable on $D$ and $f \geq g$ on $D$, then

$$\iint_D f(x, y) d\sigma \geq \iint_D g(x, y) d\sigma. \tag{3.2.11}$$

**Proof.** If $f, g$ are integrable on $D$, then $f - g$ is also integrable on $D$. As $f - g \geq 0$, by property **3** *a)* it follows that $\iint_D (f - g) d\sigma \geq 0$, whence, due to the property of linearity,

$$\iint_D f(x, y) d\sigma \geq \iint_D g(x, y) d\sigma. \; \blacksquare \tag{3.2.12}$$

*c)* If $|f|$ is integrable on $D$, then

$$\left| \iint_D f(x, y) d\sigma \right| \leq \iint_D |f(x, y)| d\sigma. \tag{3.2.13}$$

**Proof.** Obviously, $\pm f \leq |f|$; if we integrate this inequality and if we take into account the property **3 b)**, it follows that

$$\pm \iint_D f(x, y)\,d\sigma \leq \iint_D |f(x, y)|\,d\sigma. \blacksquare \qquad (3.2.14)$$

**4. PROPERTIES OF AVERAGE**

**Theorem 3.3.** *Let* $f : D \to \Re$ *be integrable on the compact domain* $D \subseteq \Re^2$. *Let*

$$M = \sup_{(x, y) \in D} f(x, y), \qquad m = \inf_{(x, y) \in D} f(x, y). \qquad (3.2.15)$$

*Then*

*i) There exists* $\lambda \in \Re$, $m < \lambda < M$, *such that*

$$\iint_D f(x, y)\,d\sigma = \lambda |D|. \qquad (3.2.16)$$

*ii) If* $f \in C^1(D)$, *then one can find a point* $(\alpha, \beta) \in D$, *such that*

$$\iint_D f(x, y)\,d\sigma = f(\alpha, \beta) |D|. \qquad (3.2.17)$$

*Remark.* The point *ii)* is, in fact, Lagrange's theorem, generalized for the double integral.

**Proof.**

*i)* $\to$ From the hypothesis, we have

$$m \leq f(x, y) \leq M \text{ for any } (x, y) \in D. \qquad (3.2.18)$$

From the monotonicity property *3b)*, it follows that

$$m \iint_D d\sigma \leq \iint_D f(x, y)\,d\sigma \leq M \iint_D d\sigma, \qquad (3.2.19)$$

therefore, due to the property of linearity, we have

$$m \underbrace{\iint_D \mathrm{d}\sigma}_{=|D|} \le \iint_D f(x, y)\mathrm{d}\sigma \le M \underbrace{\iint_D \mathrm{d}\sigma}_{=|D|}. \tag{3.2.20}$$

There are two possibilities:

1. $|D| = 0$ and then $\iint_D f(x, y)\mathrm{d}\sigma = 0$ (according to property

1), hence $\lambda = 0$.

2. $|D| \ne 0$; then we divide the relation (3.2.20) by $|D|$ and we

get

$$m \le \frac{\displaystyle\iint_D f(x, y)\mathrm{d}\sigma}{|D|} \le M, \quad \text{therefore} \quad \lambda = \frac{\displaystyle\iint_D f(x, y)\mathrm{d}\sigma}{|D|}. \tag{3.2.21}$$

*ii)* $\rightarrow$ If, moreover, $f \in C^0(D)$, as $D$ is compact, it follows

that there exists a point $(\alpha, \beta) \in D$, such that $f(\alpha, \beta) = \lambda$ (this is a

generalization of Darboux's property for functions of a real variable).

By replacing this value in (3.2.16), we obtain (3.2.17). ◻

## 3.3. HOW TO COMPUTE A DOUBLE INTEGRAL

We shall compute the double integral taking into account the

degree of complexity of the domain: firstly, on rectangles and then, on

domains simple with respect to the axes of coordinates.

## 3.3.1. THE CALCULUS OF THE DOUBLE INTEGRAL ON A RECTANGLE

Consider the rectangle $D = [a, b] \times [c, d]$ and let $f : D \subseteq \mathfrak{R}^2 \to \mathfrak{R}$ be an integrable on $D$ function.

We consider the partition $\Delta$ formed of rectangles, more precisely, $\Delta = \{ D_{ij}, \ i = \overline{1, n}, \ j = \overline{1, m} \}$(figure 3.6), created by using the points:

$$a = x_0 < x_1 < x_2 < ... < x_{i-1} < x_i < ... < x_{n-1} < x_n = b,$$
$$c = y_0 < y_1 < y_2 < ... < y_{j-1} < y_j < ... < y_{m-1} < y_m = d. \qquad (3.3.1)$$

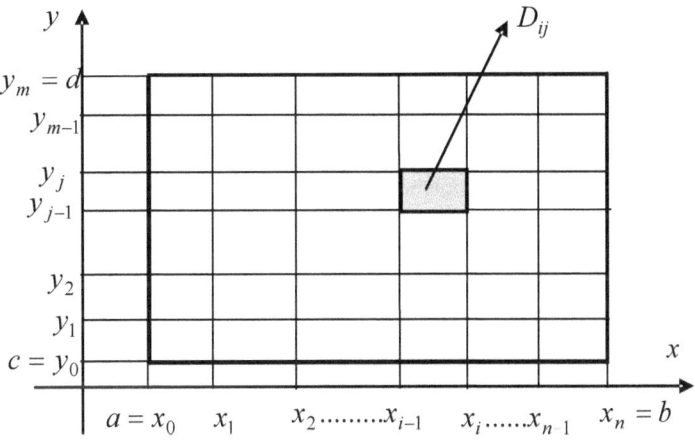

*Figure 3. 6. The partition for the calculus of the double integral on a rectangle*

Let

$$M_{ij} = \sup_{(x, y) \in D_{ij}} f(x, y), \quad m_{ij} = \inf_{(x, y) \in D_{ij}} f(x, y). \qquad (3.3.2)$$

The following inequality holds true on each $D_{ij}$:

$$m_{ij} \le f(x, y) \le M_{ij}, \quad (x, y) \in D_{ij}. \qquad (3.3.3)$$

Let $F(x) = \int_c^d f(x, y)\,dy$, where the integral is taken with respect to $y$, having $x$ as a parameter. If $\xi_i \in [x_{i-1}, x_i]$, then

$$m_{ij} \leq f(\xi_i, y) \leq M_{ij}, \quad (\xi_i, y) \in D_{ij}. \tag{3.3.4}$$

Integrating between $y_{j-1}$, $y_j$, we obtain

$$m_{ij}(y_j - y_{j-1}) \leq \int_{y_{j-1}}^{y_j} f(\xi_i, y)\,dy \leq M_{ij}(y_j - y_{j-1}). \tag{3.3.5}$$

We now sum over $j$, thus getting

$$\sum_{j-1}^{m} m_{ij}(y_j - y_{j-1}) \leq \sum_{j=1}^{m} \int_{y_{j-1}}^{y_j} f(\xi_i, y)\,dy \leq \sum_{j-1}^{m} M_{ij}(y_j - y_{j-1}); \tag{3.3.6}$$

but

$$\sum_{j=1}^{m} \int_{y_{j-1}}^{y_j} f(\xi_i, y)\,dy = \int_c^d f(\xi_i, y)\,dy \equiv F(\xi_i). \tag{3.3.7}$$

This yields

$$\sum_{j=1}^{m} m_{ij}(y_j - y_{j-1}) \leq F(\xi_i) \leq \sum_{j=1}^{m} M_{ij}(y_j - y_{j-1}), \quad i = \overline{1, n}. \tag{3.3.8}$$

We multiply by $(x_i - x_{i-1})$, summing again, this time over $i$.
We have

$$\sum_{i=1}^{n} \sum_{j=1}^{m} m_{ij}(y_j - y_{j-1})(x_i - x_{i-1}) \leq \underbrace{\sum_{i=1}^{n} F(\xi_i)(x_i - x_{i-1})}_{\sigma_\delta(F, \xi_i)} \leq$$
$$\leq \sum_{i=1}^{n} \sum_{j=1}^{m} M_{ij}(y_j - y_{j-1})(x_i - x_{i-1}), \tag{3.3.9}$$

where $\left(y_j - y_{j-1}\right)\left(x_i - x_{i-1}\right) = \left|D_{ij}\right|$ and $\sigma_\delta\left(F, \xi_i\right)$ is the Riemann sum with respect to $F$, associated to the partition $\delta = \left\{a = x_0, x_1, \ldots, x_n = b\right\}$ and to the intermediate points $\xi_i$. But

$$\lim_{\nu(\delta) \to 0} \sigma_\delta\left(F, \xi_i\right) = \int_a^b F(x)\,\mathrm{d}x. \tag{3.3.10}$$

The above inequality can be also written as follows

$$s_\Delta(f) \le \sigma_\delta\left(F, \xi_i\right) \le S_\Delta(f), \tag{3.3.11}$$

where $s_\Delta(f)$ is the lower Darboux sum for $f(x, y)$, $\sigma_\delta\left(F, \xi_i\right)$ is the Riemann sum for $F(x)$ and $S_\Delta(f)$ is the upper Darboux sum for $f(x, y)$.

As $f(x, y)$ is integrable on $D$, $S_\Delta(f) - s_\Delta(f)$ tends to zero if $\nu(\Delta) \to 0$. Therefore, it follows that

$$\iint_D f(x, y)\,\mathrm{d}x\,\mathrm{d}y = \int_a^b F(x)\,\mathrm{d}x, \tag{3.3.12}$$

or, getting beck to the notation of $F(x)$, we have

$$\boxed{\iint_D f(x, y)\,\mathrm{d}x\,\mathrm{d}y = \int_a^b \mathrm{d}x \int_c^d f(x, y)\,\mathrm{d}y}. \tag{3.3.13}$$

*This formula reduces the calculus of the double integral to the successive calculus of two simple integrals.*

By inverting the roles of the variables, we infer same way the following formula:

$$\boxed{\iint_D f(x, y)\,\mathrm{d}x\,\mathrm{d}y = \int_c^d \mathrm{d}y \int_a^b f(x, y)\,\mathrm{d}x}. \tag{3.3.14}$$

*Example*. Compute the integral

$$J = \int_1^2 \int_1^2 \frac{1}{(x+2y)^2} \, dx \, dy. \tag{3.3.15}$$

**Solution.** The domain is $D = [1,2] \times [1,2]$, from the figure

3.7.

I. We apply formula (3.3.13) and we get

$$J = \int_1^2 \int_1^2 \frac{1}{(x+2y)^2} \, dx \, dy = \int_1^2 \left( -\frac{1}{2(x+2y)} \right) \Bigg|_{y=1}^{y=2} \, dx =$$

$$= \frac{1}{2} \int_1^2 \left( -\frac{1}{x+4} + \frac{1}{x+2} \right) \, dx = \frac{1}{2} \ln \frac{x+2}{x+4} \Bigg|_{x=1}^{x=2} = \tag{3.3.16}$$

$$= \frac{1}{2} \left( \ln \frac{2}{3} - \ln \frac{3}{5} \right),$$

therefore

$$\boxed{J = \frac{1}{2} \ln \frac{10}{9}.} \tag{3.3.17}$$

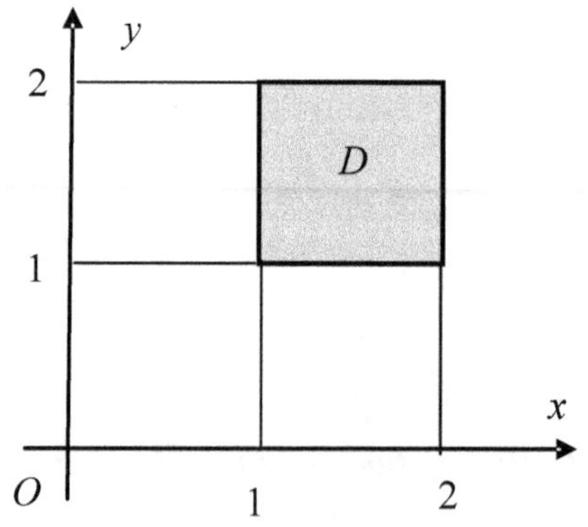

*Figure 3. 7. The calculus of the integral on the square $D = [1,2] \times [1,2]$*

II. Applying now formula (3.3.14), we obtain

144

$$J = \int_1^2 dy \int_1^2 \frac{1}{(x+2y)^2} dx = \int_1^2 -\frac{1}{x+2y}\Big|_{x=1}^{x=2} dy =$$

$$= \int_1^2 \left( -\frac{1}{2y+2} + \frac{1}{1+2y} \right) dy = -\frac{1}{2}\ln(1+y)\Big|_{y=1}^{y=2} + \qquad (3.3.18)$$

$$+\frac{1}{2}\ln(1+2y)\Big|_{y=1}^{y=2} = \frac{1}{2}(\ln 5 - \ln 3 - \ln 3 + \ln 2),$$

hence we get the same value, i.e.,

$$\boxed{J = \frac{1}{2}\ln\frac{10}{9}}. \qquad (3.3.19)$$

### *Particular case*

If $f(x, y) = p(x)q(y)$, then formula (3.3.13) becomes:

$$\iint_D p(x)q(y) dx\, dy = \int_a^b dx \int_c^d p(x)q(y) dy =$$

$$\qquad (3.3.20)$$

$$= \int_a^b p(x) dx \int_c^d q(y) dy,$$

hence the double integral becomes **the product** of two simple integrals, i.e.,

$$\boxed{\iint_D p(x)q(y) dx\, dy = \int_a^b p(x) dx \cdot \int_c^d q(y) dy}. \qquad (3.3.21)$$

*Example.* Compute the integral

$$J \equiv \iint_D x^3 y^2 \, dx\, dy, \text{ where } D = [1,3]\times[-1,1]. \qquad (3.3.22)$$

**Solution.** This is the particular case discussed above. Here,

$$f(x, y) = \underbrace{x^3}_{p(x)} \cdot \underbrace{y^2}_{q(y)}.$$

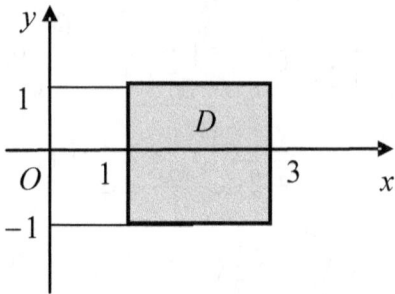

*Figure 3. 8. The calculus of the integral on the rectangle $D = [1,3] \times [-1,1]$*

We apply (3.3.21), thus getting

$$J = \int_1^3 x^3 \, dx \cdot \int_{-1}^1 y^2 \, dy = \frac{x^4}{4} \bigg|_{x=1}^{x=3} \cdot \frac{y^3}{3} \bigg|_{y=-1}^{y=1} = \left( \frac{81}{4} - \frac{1}{4} \right) \cdot \frac{2}{3}, \qquad (3.3.23)$$

hence

$$\boxed{J = \frac{40}{3}}. \qquad (3.3.24)$$

## 3.3.2. THE CALCULUS OF THE DOUBLE INTEGRAL ON DOMAINS SIMPLE $Oy$

The standard type of such a domain is shown in figure 3.9.

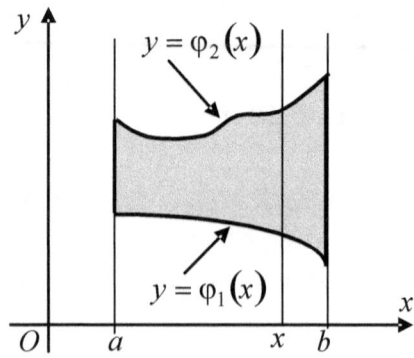

*Figure 3. 9. An example of a domain simple Oy*

Such domains are bounded to the left and to the right by two parallels to the $Oy$ axis, say, $x = a,\ x = b$.

They are limited above and below by the continuous curves $y = \varphi_1(x),\ y = \varphi_2(x)$, where $\varphi_1(x) < \varphi_2(x)$ for $x \in [a, b]$; thus, each one of the two curves crosses any parallel to $Oy$ only once.

*Examples* of domains which are **not** simple with respect to the $Oy$ axis are shown in the figure 3.10 below.

Let $D$ be a domain simple $Oy$ and $f$ an integrable on $D$ function. We wish to compute

$$J = \iint\limits_{D} f(x,\ y)\,\mathrm{d}x\,\mathrm{d}y. \tag{3.3.25}$$

We enclose the domain $D$ within the rectangle $\Omega = [a, b] \times [c, d]$ (figure 3.11) and we define the function $f^*$ as follows:

$$f^*(x,\ y) = \begin{cases} f(x,\ y), & (x,\ y) \in D, \\ 0, & (x,\ y) \in \Omega\,/\,D. \end{cases} \tag{3.3.26}$$

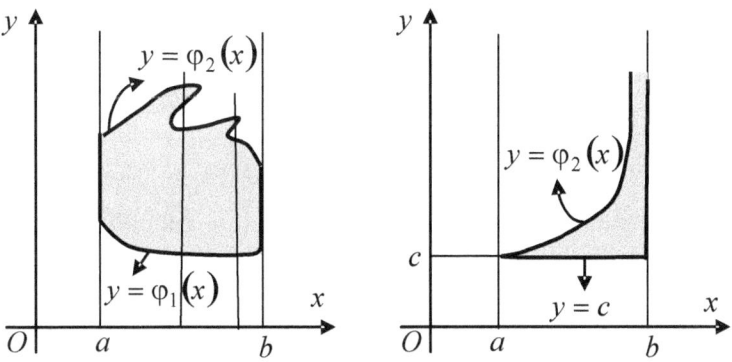

*Figure 3.10. Domains which are not simple Oy*

Therefore, we can write:

$$\iint\limits_{D} f\left(x,y\right) dx\,dy = \int\limits_{a}^{b}\int\limits_{c}^{d} f^{*}\left(x,y\right) dx\,dy =$$

$$= \int\limits_{a}^{b} dx \int\limits_{c}^{d} f^{*}\left(x,y\right) dy = \int\limits_{a}^{b} dx \int\limits_{\varphi_{1}(x)}^{\varphi_{2}(x)} f\left(x,y\right) dy, \qquad (3.3.27)$$

whence we infer **the formula for the calculus of the double integral on domains simple** *Oy*:

$$\boxed{\iint\limits_{D} f\left(x,y\right) dx\,dy = \int\limits_{a}^{b} dx \int\limits_{y=\varphi_{1}(x)}^{y=\varphi_{2}(x)} f\left(x,y\right) dy}. \qquad (3.3.28)$$

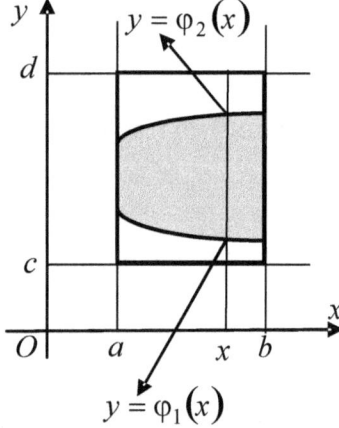

*Figure 3.11. The enclosure of the domain D within the rectangle* $\Omega = [a,b] \times [c,d]$

*Remark.* Let us consider the problem of finding the area of a domain contained between the graph of a function *f*, defined and continuous on a real interval $[a,b]$, and the *Ox* axis. (figure 3.12).

This domain is obviously simple *Oy*. Therefore, the hatched area is calculated as follows:

$$|D| = \iint_D dx\, dy = \int_a^b dx \int_{y=0}^{y=f(x)} dy = \int_a^b (y)\Big|_{y=0}^{y=f(x)} dx =$$

$$= \int_a^b f(x)dx,$$
(3.3.29)

and we obtain the well-known formula (from high school)

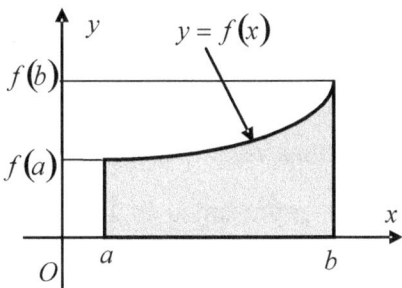

*Figure 3.12. The domain contained between the graph of a function and the Ox axis*

$$\boxed{|D| = \int_a^b f(x)dx}.$$
(3.3.30)

*Example.* Compute the following integral

$$J = \iint_D (2y - x)dx\, dy,$$
(3.3.31)

where $D$ is the domain bounded by the curves $C_1 : y = 2 - x^2$, $C_2 : y = x$ (figure 3.13).

**Solution.** We notice that $C_1$ is a parabola and $C_2$ is the first bisector. We shall cross the two curves, by solving the system

$$\begin{cases} y = 2 - x^2 \\ y = x \end{cases} \Rightarrow x = 2 - x^2 \Rightarrow x^2 + x - 2 = 0 \Rightarrow$$
(3.3.32)

$$\Rightarrow \begin{cases} x_1 = 1,\ y_1 = 1 \\ x_2 = -2,\ y_2 = -2. \end{cases}$$

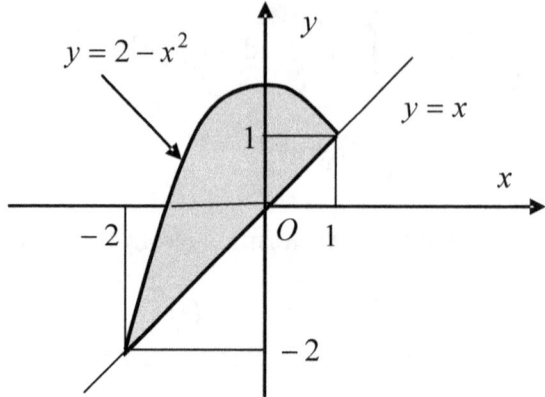

*Figure 3.13. The domain simple Oy, bounded by the curves* $y = 2 - x^2$, $y = x$

The domain $D$ represented in figure 3.13 is simple $Oy$. We have $\varphi_1(x) = x$, $\varphi_2(x) = 2 - x^2$. Therefore, applying formula (3.3.28), we find, step by step,

$$J = \int\limits_{-2}^{1} dx \int\limits_{y=x}^{y=2-x^2} (2y - x)dy = \int\limits_{-2}^{1} \left(y^2 - xy\right)\Big|_{y=x}^{y=2-x^2} dx =$$

$$= \int\limits_{-2}^{1} \left[\left(2 - x^2\right)^2 - x\left(2 - x^2\right)\right] dx + \int\limits_{-2}^{1} \left(x^2 - x^2\right)dx =$$

$$= \int\limits_{-2}^{1} \left(4 - 4x^2 + x^4 + x^3 - 2x\right)dx = \tag{3.3.33}$$

$$= \left(4x - \frac{4x^3}{3} + \frac{x^5}{5} + \frac{x^4}{4} - x^2\right)\Big|_{x=-2}^{x=1}$$

$$\Rightarrow \boxed{J = \frac{107}{20}}.$$

### 3.3.3. THE CALCULUS OF THE DOUBLE INTEGRAL ON DOMAINS SIMPLE $Ox$

The standard domain simple $Ox$ is bounded above and below by the parallels to $Ox$, $y = c$, $y = d$ and to the right and to the left it is limited by the curves $x = \psi_1(y)$ and $x = \psi_2(y)$, defined for $y \in [c, d]$.

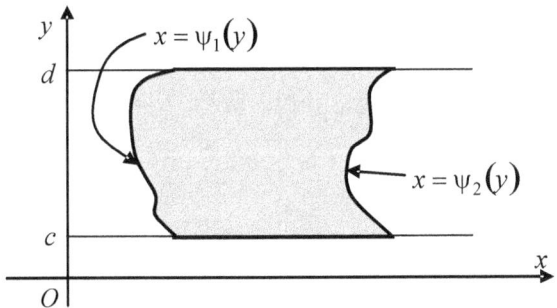

*Figure 3.14. The standard domain simple Ox*

Any parallel to $Ox$ crosses each one of these curves just once.

An example of a simple $Ox$ domain is given in figure 3.14.

*Examples* of domains which are **not** simple $Ox$ :

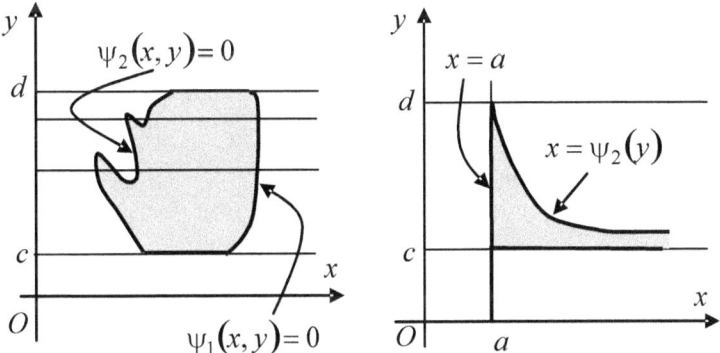

*Figure 3.15. Domains which are not simple Ox*

We can prove a formula of calculation for the double integrals on domains simple $Ox$ sameway as for domains simple $Oy$, i.e., by using formula (1.3.2). We obtain

$$\iint_D f(x, y)\,dx\,dy = \int_c^d dy \int_{x=\psi_1(y)}^{x=\psi_2(y)} f(x, y)\,dx. \qquad (3.3.34)$$

*Example*: Compute the area of the domain bounded by the curves $y = \cos x$, $y = \cos 2x$, $y = 0$.

The domain represented in figure 3.16 is simple $Ox$. It is bounded by the horizontal straight lines $y = 0$, $y = 1$ and to the right and to the left, respectively, it is bounded by the curves

$$x = \arccos y, \qquad x = \frac{1}{2}\arccos y.$$

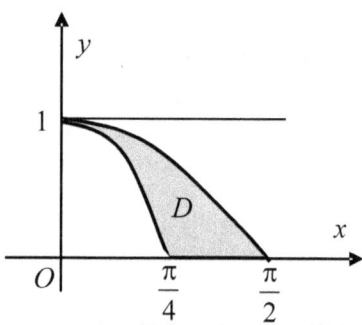

Figure 3.16. The domain simple Ox bounded by the curves
$y = \cos x$, $y = \cos 2x$, $y = 0$

The area of the domain $D$ is

$$|D| = \iint_D dx\,dy, \qquad (3.3.35)$$

and it is computed using the previous formula, in which

$$c = 0, \quad d = 1, \quad \psi_1(y) = \frac{1}{2}\arccos y, \quad \psi_2(y) = \arccos y. \qquad (3.3.36)$$

Therefore

$$|D| = \int_0^1 dy \int_{\frac{\arccos y}{2}}^{\arccos y} dx = \int_0^1 \left( \arccos y - \frac{1}{2}\arccos y \right) dy =$$

$$= \frac{1}{2}\int_0^1 \arccos y \, dy =$$

$$= \frac{1}{2} y \arccos y \Big|_0^1 - \frac{1}{2}\int_0^1 y \cdot \frac{-1}{\sqrt{1-y^2}} dy =$$

$$= 0 + \frac{1}{2}\left(-\sqrt{1-y^2}\right)\Big|_{y=0}^{y=1} \Rightarrow \boxed{|D| = \frac{1}{2}}.$$

(3.3.37)

## 3.4. CHANGES OF VARIABLES IN THE DOUBLE INTEGRAL

Consider the coordinate planes $uov$ and $xOy$, and let $\mathcal{D} \subset uov$ be a compact domain (closed and bounded). The boundary of $\mathcal{D}$, denoted by $\gamma$, is represented by the equations

$$fr\,\mathcal{D} \equiv \gamma : \begin{cases} u = u(t) \\ v = v(t) \end{cases}, \quad t \in [a, b] \subset \mathfrak{R}.$$ (3.4.1)

We consider the transformation **T**, defined as follows

$$\mathbf{T} : \begin{cases} x = x(u, v) \\ y = y(u, v) \end{cases}, \quad (u,v) \in \mathcal{D}, \quad \mathbf{T} \in C^1(\mathcal{D}),$$ (3.4.2)

which maps $\mathcal{D}$ into the compact domain $D \subset (xOy)$.

**T** maps the boundary $\gamma$ of $D$ into the boundary $\Gamma$ of $D$. We assume that **T** is nonsingular (regular), which means that

153

$$\frac{D(x, y)}{D(u, v)} \equiv \begin{vmatrix} \dfrac{\partial x}{\partial u} & \dfrac{\partial x}{\partial v} \\ \dfrac{\partial y}{\partial u} & \dfrac{\partial y}{\partial v} \end{vmatrix} \neq 0, \quad (u, v) \in \mathcal{D}. \tag{3.4.3}$$

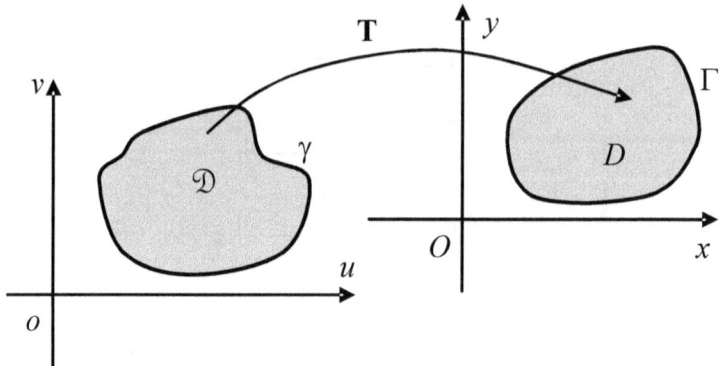

Figure 3. 17. The transformation which maps $\mathcal{D}$ into D

Let us take now a function $f : D \to \mathfrak{R}$, continuous on D (i.e., $f \in C^0(D)$). The following theorem is true:

**Theorem 5.4.** (*CHANGE OF VARIABLES IN THE DOUBLE INTEGRAL*). *If* $x, y \in C^2(D)$, *then*

$$\iint_D f(x, y) \, dx \, dy = \iint_{\mathcal{D}} f\big(x(u, v), y(u, v)\big) \left| \frac{D(x, y)}{D(u, v)} \right| du \, dv. \tag{3.4.4}$$

**The schedule of the proof**

**1)** We show that the element of area $dx \, dy$ is transformed by **T** as follows:

$$dx \, dy = \left| \frac{D(x, y)}{D(u, v)} \right| du \, dv, \tag{3.4.5}$$

firstly proving that there exists a point $(u_0, v_0) \in \mathcal{D}$ such that

$$|D| = \left.\frac{D(x,y)}{D(u,v)}\right|_{(u_0, v_0)} |\mathcal{D}|. \tag{3.4.6}$$

**2)** We consider the Riemann sum on $D$:

$$\sum_{i=1}^{n} f(x_i, y_i)|D_i|, \quad D_i \in \Delta \equiv \{D_1, D_2, \ldots, D_n\}, \tag{3.4.7}$$

for the partition $\Delta$ on $D$.

**3)** In (3.4.7) we apply (3.4.6) for each $D_i$ and we replace $x_i$, $y_i$ by their expressions, in $u_i$, $v_i$. We obtain the Riemann sum for the function $f(x(u,v), y(u,v))\left|\dfrac{D(x,y)}{D(u,v)}\right|$ with respect to the variables $u$, $v$.

### 3.4.1. PARTICULAR CASE: POLAR COORDINATES

In the case of polar coordinates (see [2,4,7,10]),

$$\mathbf{T}: \begin{cases} x = \rho\cos\theta \\ y = \rho\sin\theta \end{cases}. \tag{3.4.8}$$

The Jacobian of the transformation is

$$\frac{D(x,y)}{D(\rho,\theta)} \equiv \begin{vmatrix} \dfrac{\partial x}{\partial \rho} & \dfrac{\partial x}{\partial \theta} \\ \dfrac{\partial y}{\partial \rho} & \dfrac{\partial y}{\partial \theta} \end{vmatrix} = \begin{vmatrix} \cos\theta & -\rho\sin\theta \\ \sin\theta & \rho\cos\theta \end{vmatrix}, \tag{3.4.9}$$

such that

$$\frac{D(x,y)}{D(\rho,\theta)} = \rho \geq 0. \tag{3.4.10}$$

Therefore

$$\iint_D f(x, y)\,dx\,dy = \iint_{\mathcal{D}} f(\rho\cos\theta, \rho\sin\theta x)\rho\,d\rho\,d\theta.$$  (3.4.11)

*Example*. Compute the area of the plane domain bounded by the curves $xy = 1$, $xy = 2$, $y = x$, $y = 3x$.

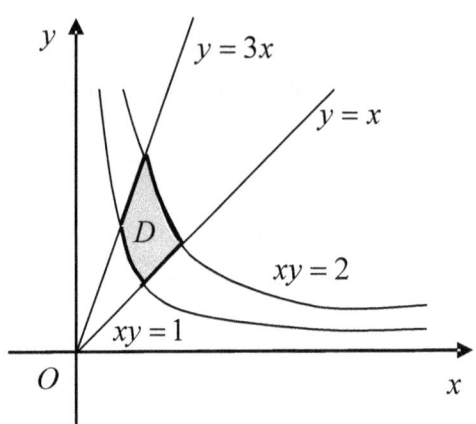

*Figure 3. 18. The area of the domain bounded by the curves*
*$xy = 1, xy = 2, y = x, y = 3x$*

**Solution.** We notice that $xy = 1$, $xy = 2$ are equations of hyperbolae, $y = x$ is the first bisector, and $y = 3x$ is a straight line which passes through the origin and has the slope equal to 3 (figure 3.18).

The area of $D$ is given by

$$|D| = \iint_D dx\,dy.$$  (3.4.12)

We make the change $\begin{cases} xy = u \\ \dfrac{x}{y} = v \end{cases} \Rightarrow \mathbf{T}: \begin{cases} x = \sqrt{\dfrac{u}{v}} \\ y = \sqrt{uv} \end{cases}.$

156

The hyperbolas $xy = 1$, $xy = 2$ become the parallel straight lines $u = 1$, $u = 2$, and the straight lines $y = x$, $y = 3x$ become $v = 1$, $v = 3$, respectively. The domain $D$ turns into the rectangle $\mathcal{D}$ in the plane $uov$ (figure 3.19).

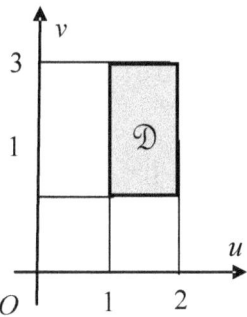

*Figure 3. 19. The domain D was mapped into the rectangle $\mathcal{D} = [1,2] \times [1,3]$*

We calculate the Jacobian

$$\frac{D(x,y)}{D(u,v)} = \begin{vmatrix} \dfrac{\partial x}{\partial u} & \dfrac{\partial x}{\partial v} \\ \dfrac{\partial y}{\partial u} & \dfrac{\partial y}{\partial v} \end{vmatrix} = \begin{vmatrix} \dfrac{1}{2\sqrt{uv}} & -\dfrac{\sqrt{u}}{2v\sqrt{v}} \\ \dfrac{1}{2}\sqrt{\dfrac{v}{u}} & \dfrac{1}{2}\sqrt{\dfrac{u}{v}} \end{vmatrix}, \qquad (3.4.13)$$

where we previously computed

$$\frac{\partial x}{\partial u} = \frac{\partial}{\partial u}\left( u^{\frac{1}{2}} v^{-\frac{1}{2}} \right) = \frac{1}{2} u^{-\frac{1}{2}} v^{-\frac{1}{2}} = \frac{1}{2\sqrt{uv}},$$

$$\frac{\partial x}{\partial v} = \frac{\partial}{\partial v}\left( u^{\frac{1}{2}} v^{-\frac{1}{2}} \right) = -\frac{1}{2} u^{\frac{1}{2}} v^{-\frac{3}{2}} = -\frac{\sqrt{u}}{2v\sqrt{v}},$$

$$\frac{\partial y}{\partial u} = \frac{\partial}{\partial u}\left( u^{\frac{1}{2}} v^{\frac{1}{2}} \right) = \frac{1}{2} u^{-\frac{1}{2}} v^{\frac{1}{2}} = \frac{1}{2}\sqrt{\frac{v}{u}}, \qquad (3.4.14)$$

$$\frac{\partial y}{\partial v} = \frac{\partial}{\partial v}\left( u^{\frac{1}{2}} v^{\frac{1}{2}} \right) = \frac{1}{2}\sqrt{\frac{u}{v}}.$$

We get

$$\frac{D(x, y)}{D(u, v)} = \frac{1}{4}\left(\frac{1}{\sqrt{uv}} \cdot \sqrt{\frac{u}{v}} + \sqrt{\frac{v}{u}} \cdot \frac{\sqrt{u}}{2v\sqrt{v}}\right) = \frac{1}{2v} \neq 0, \qquad (3.4.15)$$

because $v \neq 0$ in $\mathcal{D}$. Also, the Jacobian is positive in $\mathcal{D}$, because $v > 0$.
It follows that

$$|D| = \iint_D dx\,dy = \iint_{\mathcal{D}} \frac{1}{2v} du\,dv. \qquad (3.4.16)$$

The integrand is in the particular case $\varphi(u, v) = p(v) \cdot 1$,
therefore

$$|D| = \int_1^2 du \cdot \int_1^3 \frac{1}{2v} dv = u\Big|_1^2 \cdot \frac{1}{2} \ln v\Big|_{v=1}^{v=3} \Rightarrow \boxed{|D| = \frac{1}{2}\ln 3}. \qquad (3.4.17)$$

# 3.5. APPLICATIONS OF THE DOUBLE INTEGRAL IN MECHANICS AND GEOMETRY

To do such applications, we must regard plane domains as plates. The plates are still tridimensional bodies, but they have only two significant dimensions, one of them (e.g., the height) being much lesser that the others. With this agreement, we can compute by using the double integral the following physical quantities:

**1. The area** of a plate (plane domain) $D$ is

$$|D| = \iint_D dx\,dy. \qquad (3.5.1)$$

**2. The mass** of a plate (plane domain) $D$ is

$$m_D = \iint_D \rho(x, y)\,dx\,dy, \qquad (3.5.2)$$

where $\rho(x, y)$ is the suface density, i.e., the mass per unit area of the domain $(\rho > 0)$.

**3. The center of mass** of a plate (plane domain) $D$ has the following coordinates

$$\bar{x} = \underbrace{\frac{\iint\limits_{D} x\rho(x, y)\,d\,x\,d\,y}{\iint\limits_{D} \rho(x, y)\,d\,x\,d\,y}}_{m_D}, \quad \bar{y} = \underbrace{\frac{\iint\limits_{D} y\rho(x, y)\,d\,x\,d\,y}{\iint\limits_{D} \rho(x, y)\,d\,x\,d\,y}}_{m_D}. \qquad (3.5.3)$$

**4. The geometric center of mass** of a plate $D$ has the following coordinates

$$\bar{x} = \underbrace{\frac{\iint\limits_{D} x\,d\,x\,d\,y}{\iint\limits_{D} d\,x\,d\,y}}_{|D|}, \quad \bar{y} = \underbrace{\frac{\iint\limits_{D} y\rho(x, y)\,d\,x\,d\,y}{\iint\limits_{D} \rho(x, y)\,d\,x\,d\,y}}_{|D|}. \qquad (3.5.4)$$

If the domain is homogeneous, this meaning that $\rho(x, y) = k$, with $k$ a positive constant, then

$$\bar{x} = \frac{\iint\limits_{D} x k\,d\,x\,d\,y}{\iint\limits_{D} k\,d\,x\,d\,y} = \frac{k\iint\limits_{D} x\,d\,x\,d\,y}{k\iint\limits_{D} d\,x\,d\,y} = \bar{\bar{x}}, \quad \bar{y} = \bar{\bar{y}}, \qquad (3.5.5)$$

accordingly.

Consequently,

**The center of mass of a homogeneous plate coincides with its geometric center of mass.**

**5. The moments of inertia** of plates (plane domains).

Consider a system composed of $n$ material points $P_k$, of masses $m_k$ respectively.

The **moment of inertia** of this system with respect to an axis or to a point is defined by the expression

$$I = \sum_{k=1}^{n} m_k d_k^2 ,$$
(3.5.6)

where $d_k$ is the distance from $P_k$ to the point or to the axis.

For a better understanding, let us consider the case $n = 3$. The corresponding system, as well as the distances of its points to a point $O$ and to a straight line $d$ are represented in figure 3.20.

The case *a)* corresponds to the moment of inertia with respect to the point $O$,

$$I_O = m_1 d_1^2 + m_2 d_2^2 + m_3 d_3^2$$
(3.5.7)

and in the case *b)* we have a moment of inertia with respect to the straight line $d$,

$$I_d = m_1 d_1'^2 + m_2 d_2'^2 + m_3 d_3'^2 .$$
(3.5.8)

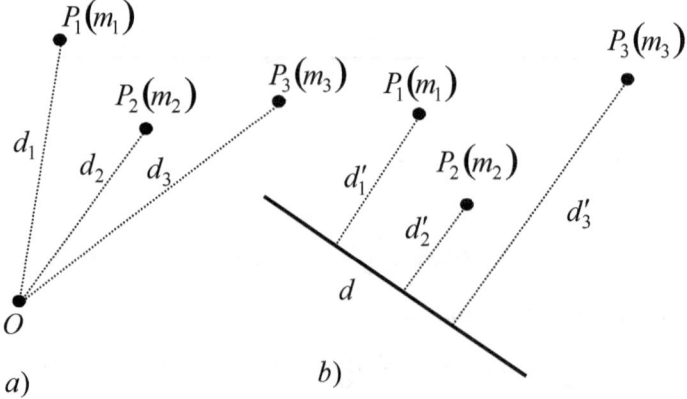

*Figure 3. 20. Moments of inertia with respect to: a) a point O, b) a straight line d*

If the system has a continuous distribution of points, i.e., if it forms a domain (for example, a plate), then the definition is still valid, but the sum turns into an integral.

a) **The moment of inertia with respect to the** $Ox$ **axis**, i.e., $I_{Ox}$. The distance from the current point to $Ox$ is $y$ (figure 3.21). Hence

$$I_x = \iint_D y^2 \rho(x, y)\, dx\, dy.$$  (3.5.9)

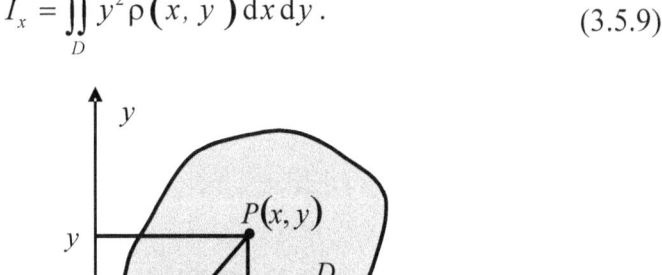

Figure 3. 21. The distances from the current point $P(x, y)$ to the axes and to the origin

b) **The moment of inertia with respect to the** $Oy$ **axis**, i.e. $I_{Oy}$. The distance from the current point $P(x, y)$ from $D$ to the $Oy$ axis is $x$. Hence,

$$I_y = \iint_D x^2 \rho(x, y)\, dx\, dy.$$  (3.5.10)

c) **The polar moment of inertia** (with respect to the origin $O$) is defined as

$$I_O = \iint_D (x^2 + y^2)\rho(x, y)\, dx\, dy,$$  (3.5.11)

161

where the distance from the current point to the origin $O$ is

$$OP = \sqrt{x^2 + y^2}.$$

We notice that

$$I_O = I_x + I_y. \qquad (3.5.12)$$

***The geometric moments of inertia*** are analogously defined, by taking $\rho = 1$.

**6. *The statical moments*** with respect to the $Ox, Oy$ axes are defined as:

$$M_x = \iint_D y \rho(x, y) \, dx \, dy, \quad M_y = \iint_D x \rho(x, y) \, dx \, dy, \qquad (3.5.13)$$

where $\rho(x, y)$ is the surface density of the domain $D$.

We notice that $\bar{x} = \dfrac{M_y}{m_D}$, $\bar{y} = \dfrac{M_x}{m_D}$, where $m_D$ is the mass of the domain (plate).

*Examples*:

**1.** Find the mass of a plane domain of the form of a quarter of the circle of radius $a$, if its surface density is $\gamma(x, y) = \sqrt{x^2 + y^2}$.

**Solution.** The mass of the domain is given by

$$m_D = \iint_D \gamma(x, y) \, dx \, dy = \iint_D \sqrt{x^2 + x^2} \, dx \, dy. \qquad (3.5.14)$$

To compute this double integral, we notice that the domain has circular symmetry; thus, it is easier to apply polar coordinates.

We obviously have $\begin{cases} x = \rho \cos\theta \\ y = \rho \sin\theta \end{cases}$, $\rho \in [0, a]$, $\theta \in \left[0, \dfrac{\pi}{2}\right]$.

The circle is of equation $x^2 + y^2 = a^2$, which yields $\rho = a$, for $\theta \in \left[0, \dfrac{\pi}{2}\right]$.

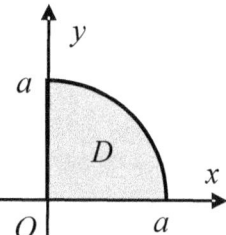

*Figure 3. 22. The plane domain of the form of a quarter of a circle*

Thus, in the plane $\rho o \theta$, the domain $D$ is mapped into the rectangle $\mathcal{D} = [0, a] \times \left[0, \dfrac{\pi}{2}\right]$ (figure 3.23).

We have $\sqrt{x^2 + y^2} = \rho$, and $\left| \dfrac{D(x, y)}{D(\rho, \theta)} \right| = \rho$. It follows that

$$m_D = \iint_D \rho \cdot \rho \, d\rho \, d\theta. \qquad (3.5.15)$$

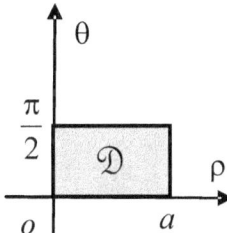

*Figure 3. 23. The transformation of the domain into the rectangle*

$$\mathcal{D} = [0, a] \times \left[0, \dfrac{\pi}{2}\right]$$

We see that the variables are separated, hence

$$m_D = \int_0^a \rho^2 \, d\rho \cdot \int_0^{\frac{\pi}{2}} d\theta = \left. \dfrac{\rho^3}{3} \right|_0^a \cdot \left. \theta \right|_0^{\frac{\pi}{2}} \Rightarrow \boxed{m_D = \dfrac{\pi a^3}{6}}. \qquad (3.5.16)$$

**2.** Find the geometric polar moment of inertia of the plate (plane domain) bounded by the rightlines $\frac{x}{a} + \frac{y}{b} = 1, \ y = 0, \ x = 0$.

**Solution.** We have

$$I_O = \iint_D \left( x^2 + y^2 \right) \cdot 1 \cdot dx\, dy. \tag{3.5.17}$$

$D$ is simple $Ox$, as well as simple $Oy$ (figure 3.24).

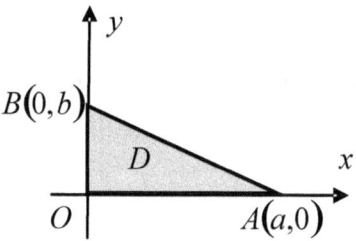

*Figure 3. 24. Plane domain bounded by the straight lines* $\frac{x}{a} + \frac{y}{b} = 1, y = 0, x = 0$

Taking into account that $D$ is simple $Oy$, it follows that

$$I_O = \int_0^a dx \int_0^{b\left(1-\frac{x}{a}\right)} \left( x^2 + y^2 \right) dy = \int_0^a [x^2 y \big|_0^{b\left(1-\frac{x}{a}\right)} + \frac{y^3}{3} \big|_0^{b\left(1-\frac{x}{a}\right)}]\, dx =$$

$$= \int_0^a x^2 \cdot b\,(1 - \frac{x}{a})\, dx + \int_0^a \frac{b^3}{3} \underbrace{(1 - \frac{x}{a})^3}_{t}\, dx. \tag{3.5.18}$$

In the last integral, we make the change of variable

$$t = 1 - \frac{x}{a} \quad \Rightarrow \quad dt = -\frac{dx}{a}. \tag{3.5.19}$$

Finally, we obtain

$$I_O = b \frac{x^3}{3} \Big|_0^a - \frac{b}{a} \cdot \frac{x^4}{4} \Big|_0^a + \frac{b^3}{3} \int_0^1 t^3 \cdot (-a\, dt) =$$

$$= \frac{ba^3}{3} - \frac{ba^3}{4} + \frac{b^3 a}{3} \cdot \frac{1}{4}, \tag{3.5.20}$$

164

or

$$I_O = \frac{ab}{12}\left(a^2 + b^2\right).$$ (3.5.21)

We obtain the same result by applying the formula of calculation of double integrals on simple $Ox$ domains (**verify!**).

**3.** Find the abscissa of the geometric center of mass of the plate of the form of a domain bounded by the curves $y = x^2$, $y^2 = x$.

**Solution.** The abscissa of the center of mass is computed by the formula

$$\overline{x} = \frac{\iint\limits_D x \, dx \, dy}{|D|}.$$ (3.5.22)

From the figure 3.25 we see that the domain is both simple $Ox$ and $Oy$. Firstly, we find $|D|$, i.e., the area of $D$:

$$|D| = \iint\limits_D dxdy = \int_0^1 dx \int_{x^2}^{\sqrt{x}} dy = \int_0^1 \left(\sqrt{x} - x^2\right)dx =$$

$$= \frac{x^{\frac{3}{2}}}{\frac{3}{2}}\Bigg|_0^1 - \frac{x^3}{3}\Bigg|_0^1 = \frac{2}{3} - \frac{1}{3} \Rightarrow \boxed{|D| = \frac{1}{3}}.$$ (3.5.23)

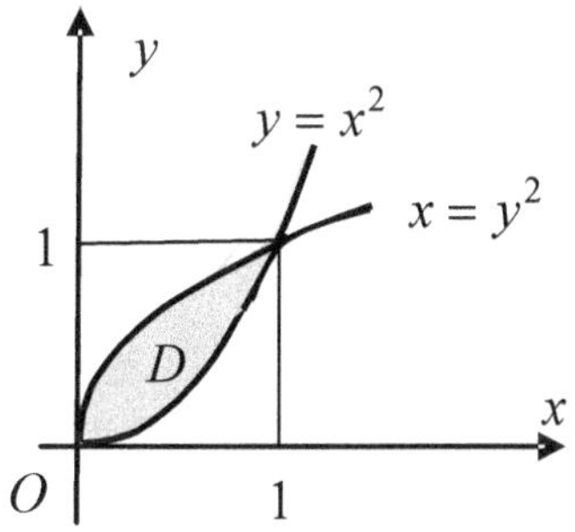

Figure 3. 25. *The domain bounded by the parabolas* $y = x^2, y^2 = x$

Then

$$M_y = \iint_D x\,dx\,dy = \int_0^1 dx \int_{x^2}^{\sqrt{x}} x\,dy = \int_0^1 \left( x\sqrt{x} - x^3 \right) dx =$$

$$= \left( \frac{x^{\frac{5}{2}}}{\frac{5}{2}} - \frac{x^4}{4} \right)\Bigg|_0^1 = \frac{2}{5} - \frac{1}{4} \Rightarrow \boxed{M_y = \frac{3}{20}}.$$

(3.5.24)

Hence,

$$\overset{=}{x} = \frac{\dfrac{3}{20}}{\dfrac{1}{3}} \Rightarrow \boxed{\overset{=}{x} = \frac{9}{20}}.$$

(3.5.25)

## 3.6. GREEN'S FORMULA

This formula establishes a connection between the second kind curvilinear integral and a particular double integral.

Let the domain $D$ be both simple $Ox$ and $Oy$ (figure 3.26).

This means that $D$ can be represented in two different manners, as follows:

*a)* As $D$ is simple $Oy$, we have

$$D = \left\{ (x,y) \in \Re^2 \,\middle|\, x \in [a,b];\ \varphi(x) \le y \le \psi(x), x \in [a,b] \right\},$$

where $\varphi, \psi \in C^0\left([a,b]\right)$.

*b)* As $D$ is simple $Ox$, it can also be represented in the form:

$$D = \left\{ (x,y) \in \Re^2 \,\middle|\, y \in [c,d];\ \alpha(y) \le x \le \beta(y), y \in [c,d] \right\},$$

where $\alpha, \beta \in C^0\left([c,d]\right)$.

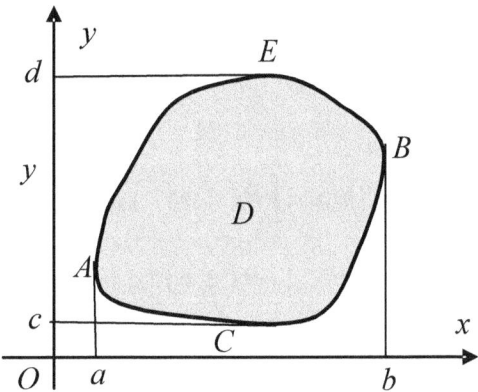

*Figure 3. 26. The domain both simple Ox and Oy for the proof of Green's formula*

Denote the boundary of $D$ by $\Gamma$, i.e. $\Gamma = fr\,D$.

**Theorem 3. 5** (**GREEN'S FORMULA**). *Let $D \subset \mathfrak{R}^2$ be a compact domain, both simple Ox and Oy. Let $P, Q \in C^1(D)$. Then the following formula is valid*

$$\oint_{\Gamma} P(x,y)\,dx + Q(x,y)\,dy = \iint_{D}\left(\frac{\partial Q}{\partial x} - \frac{\partial P}{\partial y}\right)dx\,dy\,.$$

(3.6.1)

$$\Downarrow$$

***GREEN'S FORMULA***

**\* Proof.** The boundary $\Gamma$ can be expressed in two ways, each of them emphasizing the property of $D$ to be simple $Ox$ or simple $Oy$.

I. $\Gamma = \overset{\frown}{ACB} \cup \overset{\frown}{BEA}$, where

$$\overset{\frown}{ACB}: \begin{cases} x = x \\ y = \varphi(x) \end{cases}, \quad x \in [a,b],$$

$$\overset{\frown}{AEB}: \begin{cases} x = x \\ y = \psi(x) \end{cases}, \quad x \in [a,b].$$

(3.6.2)

As $D$ is simple $Oy$, we have

$$\iint_D \frac{\partial P}{\partial y}(x,y)\,dx\,dy = \int_a^b dx \int_{\varphi(x)}^{\psi(x)} \frac{\partial P}{\partial y}(x,y)\,dy =$$

$$= \int_a^b P\big(x,\psi(x)\big)dx - \int_a^b P\big(x,\varphi(x)\big)dx =$$

$$= \int_{\overset{\frown}{AEB}} P(x,y)\,dx - \int_{\overset{\frown}{ACB}} P(x,y)\,dx = \tag{3.6.3}$$

$$= -\int_{\overset{\frown}{BEA}} P(x,y)\,dx - \int_{\overset{\frown}{ACB}} P(x,y)\,dx.$$

Therefore

$$\iint_D \frac{\partial P}{\partial y}\,dx\,dy = -\oint_\Gamma P(x,y)\,dx. \tag{3.6.4}$$

II. We can also represent the boundary in the form $\Gamma = \overset{\frown}{CBE} \cup \overset{\frown}{EAC}$, where

$$\overset{\frown}{CBE}: \begin{cases} x = \beta(y) \\ y = y \end{cases}, \quad y \in [c,d],$$

$$\overset{\frown}{CAE}: \begin{cases} x = \alpha(y) \\ y = y \end{cases}, \quad y \in [c,d]. \tag{3.6.5}$$

As $D$ is also simple $Ox$, we have

$$\iint_D \frac{\partial Q}{\partial x}(x,y)\,dx\,dy = \int_c^d dy \int_{\alpha(y)}^{\beta(y)} \frac{\partial Q}{\partial x}(x,y)\,dx =$$

$$= \int_c^d Q\big(\beta(y),y\big)dy - \int_c^d Q\big(\alpha(y),y\big)dy =$$

$$= \int_{\overset{\frown}{CBE}} Q(x,y)\,dy - \int_{\overset{\frown}{CAE}} Q(x,y)\,dy = \tag{3.6.6}$$

$$= \int_{\overset{\frown}{CBE}} Q(x,y)\,dy + \int_{\overset{\frown}{EAC}} Q(x,y)\,dy.$$

This yields

$$\iint_D \frac{\partial Q}{\partial x}\, dx\, dy = \oint_\Gamma Q(x,y)\, dx. \tag{3.6.7}$$

By subtracting the relations (3.6.7) and (3.6.4), we obtain Green's formula. ◻

*Remark.* The formula can also be applied on general domains, which can be split into domains both simple $Ox$ and $Oy$.

**PARTICULAR CASE: THE AREA OF A PLANE DOMAIN**

For

$$P(x,y) = -\frac{y}{2}, \quad Q(x,y) = \frac{x}{2}, \tag{3.6.8}$$

we infer that

$$\frac{\partial Q}{\partial x} - \frac{\partial P}{\partial y} = \frac{1}{2} - \left(-\frac{1}{2}\right) = 1 \tag{3.6.9}$$

and Green's formula becomes

$$\frac{1}{2} \oint_\Gamma x\, dy - y\, dx = \iint_D 1 \cdot dx\, dy \equiv |D|. \tag{3.6.10}$$

This means that **the area of a plane domain** can also be computed using the curvilinear integral of the second kind, i.e.

$$|D| = \frac{1}{2} \oint_\Gamma x\, dy - y\, dx, \qquad \Gamma = fr\, D, \tag{3.6.11}$$

obtained from Green's formula.

*Example.* Compute the area of the plane domain bounded by the curves $y = x^2$, $x = y^2$, $8xy = 1$.

**Solution.** The area of the domain $D$ can be computed by formula (3.6.11), for $\Gamma = \overset{\frown}{OA} \cup \overset{\frown}{AB} \cup \overset{\frown}{BO}$.

169

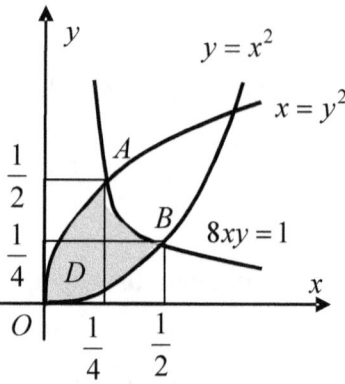

*Figure 3. 27. The domain bounded by the curves* $y = x^2, y^2 = x, 8xy = 1$

The coordinates of the points $A, B$ are determined as follows:

$$A : \begin{cases} y = x^2 \\ 8xy = 1 \end{cases} \Rightarrow 8x^3 = 1 \Rightarrow \begin{cases} x = \dfrac{1}{2} \\ y = \dfrac{1}{4} \end{cases}, \quad B : \begin{cases} x = y^2 \\ 8xy = 1 \end{cases} \Rightarrow \begin{cases} x = \dfrac{1}{4} \\ y = \dfrac{1}{2} \end{cases}. \quad (3.6.12)$$

Therefore

$$|D| = \frac{1}{2} \oint_{\Gamma} x\,d\,y - y\,d\,x =$$

$$= \frac{1}{2} \left[ \overbrace{\int_{\overset{\frown}{OA}} x\,d\,y - y\,d\,x}^{I_1} + \overbrace{\int_{\overset{\frown}{AB}} x\,d\,y - y\,d\,x}^{I_2} + \overbrace{\int_{\overset{\frown}{BO}} x\,d\,y - y\,d\,x}^{I_3} \right]. \quad (3.6.13)$$

We compute, one by one, the three curvilinear integrals. We have

$$I_1 = \int_{\overset{\frown}{OA}} x\,d\,y - y\,d\,x = \int_{0}^{\frac{1}{2}} \left( x \cdot 2x - x^2 \right) d\,x = \int_{0}^{\frac{1}{2}} x^2\,d\,x = \frac{x^3}{3} \Big|_{0}^{\frac{1}{2}}, \quad (3.6.14)$$

and we obtain

$$\boxed{I_1 = \frac{1}{24}}. \quad (3.6.15)$$

170

Analogously,

$$I_3 = \int\limits_{\overset{\frown}{BO}} x\,d\,y - y\,d\,x = \int\limits_{\frac{1}{2}}^{0} \left(y^2 - y \cdot 2y\right)d\,y =$$

$$= -\int\limits_{\frac{1}{2}}^{0} y^2\,d\,y = \int\limits_{0}^{\frac{1}{2}} y^2\,d\,y,$$

(3.6.16)

whence it follows that

$$\boxed{I_3 = \frac{1}{24}}.$$

(3.6.17)

Finally, we compute $I_2$ along the hyperbola $x\,y = \dfrac{1}{8}$.

From this equations it results that $y = \dfrac{1}{8x}$, hence

$d\,y = -\dfrac{1}{8x^2}\,d\,x$. Consequently,

$$I_2 = \int\limits_{\overset{\frown}{AB}} x\,d\,y - y\,d\,x = \int\limits_{\frac{1}{2}}^{\frac{1}{4}}\left[x\cdot\left(-\frac{1}{8x^2}\right)-\frac{1}{8x}\right]d\,x = \int\limits_{\frac{1}{4}}^{\frac{1}{2}}\frac{1}{4x}\,d\,x$$

(3.6.18)

$$\Rightarrow \boxed{I_2 = \frac{1}{4}\ln 2},$$

hence

$$|D| = \frac{1}{2}(I_1 + I_2 + I_3) = \frac{1}{2}\left(\frac{1}{24} + \frac{1}{4\ln 2} + \frac{1}{24}\right) =$$

$$= \frac{1}{2}\cdot\frac{1+3\ln 2}{12}.$$

(3.6.19)

Finally,

$$\boxed{\left| D \right| = \frac{1 + 3\ln 2}{24} \cong 0.13}. \qquad\qquad (3.6.20)$$

**APPLICATION: the area of a plane polygonal domain**

Let us consider the plane polygonal domain $D$ from the figure 3.28. We wish to find its area, given the coordinates of its edges $A_j\left(x_j, y_j\right)$ in a system of coordinates $xOy$.

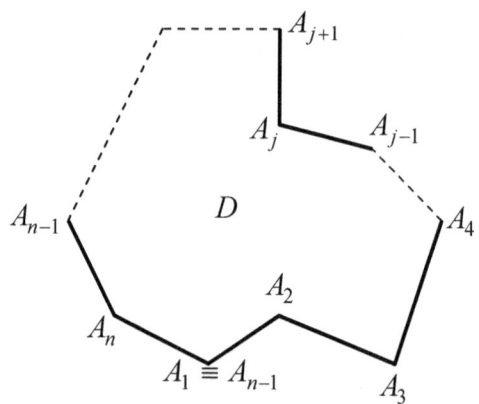

*Figure 3. 28. The polygonal domain D*

If we try to solve this problem by splitting the polygon into triangles, for example, then we will encounter major geometrical problems, because the polygon may have concavities, which should be taken into account.

In this case, Green's formula is helpful for the calculus of the area, because it can be applied without knowing the topology of the domain.

The boundary of $D$ is formed by $n$ segments, $\overline{A_j A_{j+1}}$ for $j = \overline{1, n}$. Because the polygon is closed, we denoted $A_1$ also by $A_{n+1}$.

We now apply formula (3.6.11) for the area. We get

$$|D| = \iint_D dx\, dy = \frac{1}{2} \oint_\Gamma x\, dy - y\, dx = \frac{1}{2} \int_{\underset{j=1}{\overset{n}{\bigcup}} \overline{A_j A_{j+1}}} x\, dy - y\, dx$$

$$= \frac{1}{2} \sum_{j=1}^{n} \int_{\overline{A_j A_{j+1}}} x\, dy - y\, dx,$$

(3.6.21)

in which we used the property of additivity with respect to the domain of integration of the curvilinear integral of the second kind.

We still have to compute the integrals on the segments $\overline{A_j A_{j+1}}$.

Firstly, we write the equation of the straight line passing through the points $A_j, A_{j+1}$. We have

$$y - y_j = \frac{y_{j+1} - y_j}{x_{j+1} - x_j} \left( x - x_j \right).$$

(3.6.22)

Further, using the notation

$$m_j = \frac{y_{j+1} - y_j}{x_{j+1} - x_j},$$

(3.6.23)

we obtain

$$y - y_j = m_j \left( x - x_j \right).$$

(3.6.24)

It follows that

$$d y = m_j\, d x.$$

(3.6.25)

Now, we can compute the curvilinear integral (in this case, it is clearly a *line integral!*) on the segment $\overline{A_j A_{j+1}}$:

$$\int\limits_{\overline{A_jA_{j+1}}} x\,dy - y\,dx = \int\limits_{x_j}^{x_{j+1}} \left\{ x \cdot m_j - \left[ y_j + m_j \left( x - x_j \right) \right] \right\} dx =$$

$$= \int\limits_{x_j}^{x_{j+1}} \left[ x \cdot m_j - y_j - m_j \left( x - x_j \right) \right] dx = \int\limits_{x_j}^{x_{j+1}} \left( m_j x_j - y_j \right) dx = \qquad (3.6.26)$$

$$= \left( m_j x_j - y_j \right) \int\limits_{x_j}^{x_{j+1}} dx = \left( m_j x_j - y_j \right)\left( x_{j+1} - x_j \right).$$

Taking into account the notation (3.6.23) and formula (3.6.21), we get for the area of the polygon

$$|D| = \frac{1}{2}\sum_{j=1}^{n} \int\limits_{\overline{A_jA_{j+1}}} x\,dy - y\,dx = \frac{1}{2}\sum_{j=1}^{n}\left( m_j x_j - y_j \right)\left( x_{j+1} - x_j \right) =$$

$$= \frac{1}{2}\sum_{j=1}^{n}\left( \frac{y_{j+1} - y_j}{x_{j+1} - x_j} x_j - y_j \right)\left( x_{j+1} - x_j \right) = \qquad (3.6.27)$$

$$= \frac{1}{2}\sum_{j=1}^{n}\left[ \left( y_{j+1} - y_j \right) x_j - y_j \left( x_{j+1} - x_j \right) \right],$$

and, finally,

$$\boxed{ |D| = \frac{1}{2}\sum_{j=1}^{n}\left( y_{j+1} x_j - x_{j+1} y_j \right). } \qquad (3.6.28)$$

## EXERCISES AND PROBLEMS

1. Compute the following double integrals on rectangles:

a) $I = \iint\limits_{D}(x + y)\,dx\,dy$, $D : \begin{cases} 0 \le x \le a \\ 0 \le y \le b \end{cases}$  $A : I = \dfrac{ab}{2}(a + b)$

b) $I = \iint\limits_{D}(x + 2y)\,dx\,dy$, $D : \begin{cases} -1 \le x \le 1 \\ 0 \le y \le 3 \end{cases}$  $A : I = 18$

174

c) $I = \iint_D \dfrac{dx\,dy}{(x-y)^2}$, $D : \begin{cases} 1 \le x \le 2 \\ 3 \le y \le 4 \end{cases}$     A: $I = \ln \dfrac{4}{3}$

d) $I = \iint_D x^2 y^2 \, dx\,dy$, $D : \begin{cases} -1 \le x \le 1 \\ 2 \le y \le 3 \end{cases}$     A: $I = \dfrac{38}{9}$

2. Compute the following double integrals:

a) $I = \iint_D x^2 y^3 \, dx\,dy$, $D$ being the domain bounded by the

parabola $y = x^2$ and by the straight line $y = x$.

$$A: I = \dfrac{1}{77}$$

b) $I = \iint_D xy \, dx\,dy$, $D$ being the domain bounded by the ellipse

$\dfrac{x^2}{a^2} + \dfrac{y^2}{b^2} = 1$ and by the straight lines $\dfrac{x}{a} + \dfrac{y}{b} = 1$, $x \ge 0$, $y \ge 0$.

$$A: I = \dfrac{a^2 b^2}{12}$$

c) $I = \iint_D \cos(x + y) \, dx\,dy$, $D$ being the domain bounded by

the straight lines $x = y$, $x + y = 2\pi$, $y = 0$.

$$A: I = 0$$

d) $I = \iint_D \dfrac{1}{xy} \, dx\,dy$, $D$ being the domain bounded by the

equiangular hyperbola $x y = 4$, by the parabola $y^2 = 2x$ and by the

straight line $x = a$, $a > 2$.

$$A: I = \dfrac{3}{4}\left( \ln \dfrac{a}{2} \right)^2$$

175

e) $I = \iint_D xy\,dx\,dy$, $D$ being the domain bounded by the

parabola $x = -y^2$ and by the straight lines $y = x$, $y = 1$.

$$A: I = \frac{1}{24}$$

f) $I = \iint_D xy\,dx\,dy$, $D$ being the domain bounded by the

straight lines $y = x$, $y = x - 2a$, $y = 0$, $y = a$.

$$A: I = \frac{5a^4}{3}$$

3. Compute the areas of the triangles bounded by the indicated straight lines:

a) $y = 2x - 3$, $x = 0$, $y = 0$

$A: \dfrac{9}{4}$

b) $y = -3x + 4$, $x = 0$, $y = 0$

$A: \dfrac{9}{32}$

c) $y = x + 5$, $x = 10$, $y = 5$

$A: 50$

d) $y = -4x + 6$, $x = 3$, $y = 8$

$A: 22\dfrac{1}{2}$

e) $y = x + 1$, $y = -x + 9$, $y = -\dfrac{1}{5}x + \dfrac{17}{5}$      $A: 26$

4. Compute the areas of the following plane domains, bounded by the indicated curves:

a) $y = -\sqrt{1 - x^2}$, $y = 1 - x^2$

$A: \dfrac{4}{3} + \dfrac{\pi}{2}$

b) $y = 4x - x^2$, $y^2 = 2x$ (the exterior of the parabola)

$A: \pi - \dfrac{8}{3}$

c) $y = 2 - x$, $y^2 = 4x + 4$

$A: 21\dfrac{1}{3}$

d) $x + 2y - 4 = 0$, $y^2 = 4 - x$

$A: \dfrac{4}{3}$

176

e) $y^2 = 4 - x$, $x^2 + y^2 = 4$ (the exterior of the parabola)

A: $\dfrac{32}{3} - 2\pi$

f) $3y^2 = 25x$, $5x^2 = 9y$

A: 5

5. Using appropriated changes of variables, find the following physical quantities:

### I. Using polar coordinates:

a) The area of the plate bounded by the curve $\left(x^2 + y^2\right)^2 = 2a^2\left(x^2 - y^2\right)$, $a > 0$.

$$A: |D| = \frac{\pi a^2}{4}$$

*Hint*: The transformed domain is:

$$\mathcal{D} = \left\{ (\rho, \theta),\ 0 \le \theta \le \frac{\pi}{4},\ 0 \le \rho \le \sqrt{2a^2 \cos 2\theta} \right\}.$$

b) The mass of a plate of the form of a circle of radius 1, if its surface density is $\gamma\left(x, y\right) = \left(x^2 + y^2\right)^3$.

$$A:\ m_D = \frac{\pi}{4}$$

c) The coordinates of the center of mass of the homogeneous plate bounded by the quarter of the circle $x^2 + y^2 = R^2$, $x \ge 0$, $y \ge 0$.

$$A:\ \bar{x} = \bar{y} = \frac{4R}{3\pi}$$

d) The mass of the plate bounded by the quarter of the unit circle $x^2 + y^2 = 1$, $x \ge 0$, $y \ge 0$, if its surface density is $\gamma\left(x, y\right) = \ln\left(1 + x^2 + y^2\right)$.

$$A:\ m_D = \frac{\pi}{4}\left(2\ln 2 - 1\right)$$

177

e) The mass of the plate of the form of a circular crown bounded by the circles of equations $x^2 + y^2 = 4\pi^2$, $x^2 + y^2 = 9\pi^2$, if its surface density is $\gamma(x, y) = \sin\sqrt{x^2 + y^2}$.

$$\text{A: } m_D = 10\pi^2$$

f) The coordinates of the center of mass of the plate defined at the point e).

$$\text{A: } \overline{x} = 0, \ \overline{y} = 0$$

g) The mass of the plate of surface density $\gamma(x, y) = x^2 + y^2$, bounded by the segment of circle $x^2 + y^2 - 2x - 2y = 0$, $x \geq 0$, $y \geq 0$.

$$\text{A: } m_D = 3\pi + 8$$

h) The center of mass of the nonhomogeneous plate bounded by the circle of equation $x^2 + y^2 = 2ay$, if its superficial density is $\gamma(x, y) = x^2 + y^2$.

$$\text{A: } \overline{x} = 0, \ \overline{y} = \frac{4a}{3}$$

**II. Using generalized polar coordinates** $\begin{cases} x = a\rho\cos\theta \\ y = b\rho\sin\theta \end{cases}$ :

a) The area of the plate bounded by the ellipse $\dfrac{x^2}{a^2} + \dfrac{y^2}{b^2} = 1$.

$$\text{A: } |D| = ab\pi$$

b) The area of the plate bounded by the curve $\left(\dfrac{x^2}{a^2} + \dfrac{y^2}{b^2}\right)^2 = xy$.

$$\text{A: } |D| = \frac{a^2 b^2}{2}$$

*Hint:* $\mathcal{D} = \left\{ (\rho, \theta), 0 \le \theta \le \dfrac{\pi}{2}, 0 \le \rho \le \sqrt{\dfrac{ab\sin 2\theta}{2}} \right\}$

c) The mass of the plate bounded by the ellipse $\dfrac{x^2}{a^2} + \dfrac{y^2}{b^2} = 1$, if

its surface density is $\gamma(x, y) = x^2 y^2$.

$$\text{A: } m_D = \dfrac{a^3 b^3}{24} \pi$$

6. Compute the following curvilinear integrals using Green's formula:

a) $I = \oint\limits_\Gamma (x + y)\,dx - (x - y)\,dy,\ \Gamma : \dfrac{x^2}{a^2} + \dfrac{y^2}{b^2} = 1$

$$\text{A: } I = -2\pi ab$$

b) $I = \oint\limits_\Gamma (x + y)\,dx - (x - y)\,dy,\ \Gamma : x^2 + y^2 = a^2$

$$\text{A: } I = -2\pi a^2$$

c) $I = \oint\limits_\Gamma (xy + x + y)\,dx + (xy + x - y)\,dy,\ \Gamma : \dfrac{x^2}{a^2} + \dfrac{y^2}{b^2} = 1$

$$\text{A: } I = 0$$

7. Compute the following areas by using Green's formula:

a) the area of the ellipse $\dfrac{x^2}{a^2} + \dfrac{y^2}{b^2} = 1$.

$$\text{A: } |D| = \pi ab$$

b) the area of the first cycloid loop:

$$\begin{cases} x = a(t - \sin t) \\ y = a(1 - \cos t) \end{cases}, 0 \le t \le 2\pi.$$

*Hint*: $\Gamma = \overline{OA} + \overset{\frown}{ABO}$, where $O(0,0)$, $A(2\pi,0)$, $B(\pi, 0)$.

A: $|D| = 3a^2\pi$

# Chapter 4

# THE TRIPLE INTEGRAL

## 4.1. THE DEFINITION OF THE TRIPLE INTEGRAL

Let $\Omega \subset \mathfrak{R}^3$ be a closed and bounded domain; together, these two properties are equivalent to compactness on finite-dimensional spaces. Take, also, $f : \Omega \to \mathfrak{R}$, $f = f(x, y, z)$. Consider a ***partition*** $\Delta = \{\Omega_1, \Omega_2, \ldots \Omega_n\}$ of $\Omega$ (the definition of the partition is the same as in the case of double integrals).

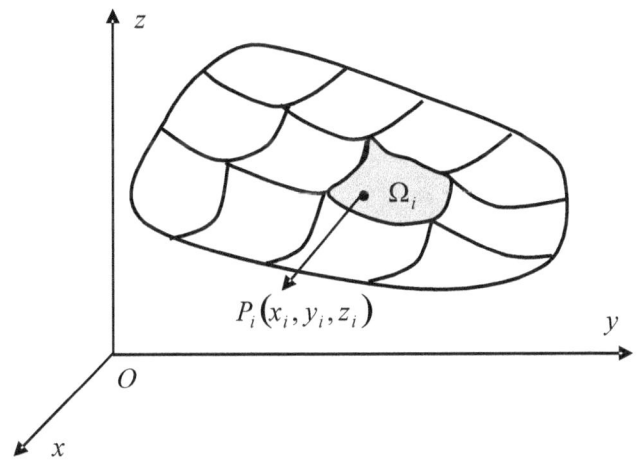

*Figure 4. 1. The partition of the domain for the triple integral*

Denote the volumes of the subdomains $\Omega_i$ by

$$\left|\Omega_i\right| = \Delta\Omega_i, \quad i = \overline{1,n}. \tag{4.1.1}$$

Then the volume of $\Omega$ is $\left|\Omega\right| = \sum_{i=1}^{n}\left|\Omega_i\right|$.

***The norm*** (mesh size) of this partition is

$$\nu(\Delta) = \max_{i=\overline{1,n}} \text{diam}\,\Omega_i, \tag{4.1.2}$$

where $\text{diam}\,\Omega_i$ is the ***diameter*** of the set $\Omega_i$.

**Definition.** Let $A \subset \mathfrak{R}^3$ and $\Gamma = fr\,A$, then

$$\text{diam}\,A = \max\left\{\left|PP'\right|, P, P' \in \Gamma\right\}. \tag{4.1.3}$$

*Examples*

1. Let $S$ be the sphere of radius $r$. Then, $\boxed{\text{diam}\,S = 2r}$.

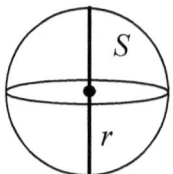

*Figure 4. 2. The diameter of a sphere*

2. For the parallelepiped $\Omega$, $\boxed{\text{diam}\,\Omega = \left|BD'\right|}$.

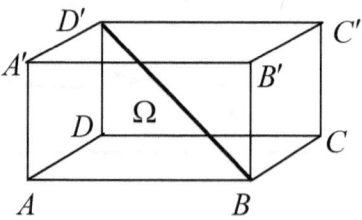

*Figure 4. 3. The diameter of a parallelepiped*

3. Let $E$ be the ellipsoid of semi-axes $a, b, c$, $2a$ being the major axis.

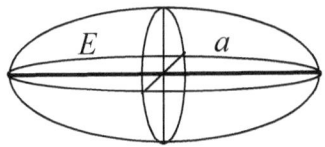

*Figure 4. 4. The diameter of an ellipsoid*

Then

$$\boxed{\operatorname{diam} E = 2a}.$$

Let $P_i\left(x_i, y_i, z_i\right) \in \Omega_i$. We set up **the Riemann sum** associated to the function $f$, to the partition $\Delta$ and to the points $P_i$,

$$\sigma_\Delta\left(f, P_i\right) = \sum_{i=1}^{n} f\left(x_i, y_i, z_i\right) \Delta\Omega_i. \tag{4.1.4}$$

**Definition.** If there exists $I \in \mathfrak{R}$, such that for any $\varepsilon > 0$ one can find a number $\eta = \eta\left(\varepsilon\right)$ with the property that $\left|\sigma_\Delta\left(f, P_i\right) - I\right| < \varepsilon$ for any partition $\Delta$ of norm $v\left(\Delta\right) < \eta$ and for any choice of the intermediate points $P_i \in \Omega_i$, then we say that $f$ is *integrable* on $\Omega$ and

$$I = \iiint_\Omega f\left(x, y, z\right) \mathrm{d}\Omega \tag{4.1.5}$$

is *the triple integral* of $f$ on $\Omega$.

In other words,

$$I = \iiint_\Omega f\left(x, y, z\right) \mathrm{d}\Omega = \lim_{\substack{v(\Delta) \to 0 \\ \forall P_i \in \Omega_i}} \sigma_\Delta\left(f, P_i\right). \tag{4.1.6}$$

*Remark.* From a logical point of view, the structure of the definition of the triple integral, as well as that of the double integral, is identical to that of the Riemann integral.

## 4.1.1. PROPERTIES OF THE TRIPLE INTEGRALS

Let us present several properties of the triple integral which result directly from their definition, based on the Riemann sum.

**1.** $\iiint\limits_{\Omega} d\Omega = |\Omega|$ (4.1.7)

is **the volume** of $\Omega$.

**Proof.** Taking $f(x, y, z) = 1$ in the definition of the integral,

we get $\sigma_{\Delta}(1, P_i) = \sum\limits_{i=1}^{n} \Delta\Omega_i$, therefore $\sigma_{\Delta}(1, P_i) = |\Omega|$, for any

partition $\Delta$ and any choice of the points $P_i$. It follows that

$$\iiint\limits_{\Omega} d\Omega = |\Omega|.$$ (4.1.8)

**2.** The triple integral is **linear**, i.e., for any $f, g$ integrable on $\Omega$ and for any $\alpha, \beta \in \Re$, we have:

$$\iiint\limits_{\Omega} (\alpha f + \beta g)(x, y, z) d\Omega =$$ (4.1.9)

$$= \alpha \iiint\limits_{\Omega} f(x, y, z) d\Omega + \beta \iiint\limits_{\Omega} g(x, y, z) d\Omega.$$

**3.** The triple integral is **additive** with respect to the domain of integration.

Indeed, consider $\Omega = \Omega_1 \cup \Omega_2$, where $\Omega_1, \Omega_2$ have, at most, boundary points in common.

Let $f$ be integrable on $\Omega_1, \Omega_2$. Then $f$ is integrable on $\Omega$ and

$$\iiint\limits_{\Omega} f(x, y, z) d\Omega = \iiint\limits_{\Omega_1} f(x, y, z) d\Omega + \iiint\limits_{\Omega_2} f(x, y, z) d\Omega.$$ (4.1.10)

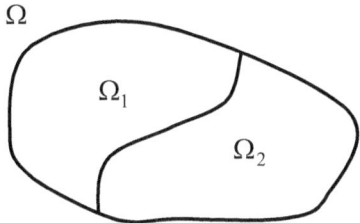

*Figure 4. 5. Domains having in common, at most, boundary points*

**4.** The triple integral is **monotone**:

**a)** If $f \geq 0$ is integrable on $\Omega$, then all the associated Riemann sums are positive, hence

$$\iiint_\Omega f(x, y, z) \, d\Omega \geq 0. \qquad (4.1.11)$$

**b)** If $f \geq g$ on $\Omega$, then

$$\iiint_\Omega f(x, y, z) \, d\Omega \geq \iiint_\Omega g(x, y, z) \, d\Omega. \qquad (4.1.12)$$

**Proof.** As $f - g \geq 0$ on $\Omega$, we can apply **4a)**, then property **2**, i.e. the property of linearity of the triple integral.

**c)** If $|f|$ is integrable on $\Omega$, then

$$\left| \iiint_\Omega f(x, y, z) \, d\Omega \right| \leq \iiint_\Omega |f(x, y, z)| \, d\Omega. \qquad (4.1.13)$$

**Proof.** We start with the obvious inequality, $\pm f \leq |f|$, which we integrate, according to **3b)**. For the left member, we apply the property of linearity of the triple integral.

## 5. Properties of average

**Theorem 4. 1.** *Consider a function* $f : \Omega \subset \mathfrak{R}^3 \to \mathfrak{R}$, *integrable on the compact domain* $\Omega$ *and let*

$$M = \sup_{(x,y,z)\in\Omega} f(x,y,z), \quad m = \inf_{(x,y,z)\in\Omega} f(x,y,z). \qquad (4.1.14)$$

*Then:*

*i) there exists* $\lambda \in (m, M)$, *such that*

$$\iiint\limits_{\Omega} f(x,y,z)\,d\Omega = \lambda|\Omega|, \qquad (4.1.15)$$

$|\Omega|$ *being the volume of* $\Omega$.

*ii) If, moreover,* $f \in C^1(\Omega)$, *then one can find a point* $(\alpha, \beta, \gamma) \in \Omega$, *such that*

$$\iiint\limits_{\Omega} f(x,y,z)\,d\Omega = f(\alpha,\beta,\gamma)|\Omega|. \qquad (4.1.16)$$

**Proof.** It is similar to that from the double integral. ◻

Darboux's integrability criterion is also valid for the triple integral.

Firstly, we consider Darboux's sums for a partition $\Delta$ of $\Omega$.
Denote by

$$m_i = \inf\{f(x,y,z)|(x,y,z)\in\Omega_i\},$$

$$\qquad (4.1.17)$$

$$M_i = \sup\{f(x,y,z)|(x,y,z)\in\Omega_i\}.$$

Then $S_\Delta(f) = \sum_{i=1}^{n} M_i \Delta\Omega_i$ is *the **upper Darboux sum*** and

$s_\Delta(f) = \sum_{i=1}^{n} m_i \Delta\Omega_i$ is *the **lower Darboux sum.***

**Theorem 4. 2.** (*DARBOUX'S INTEGRABILITY CRITERION*). *The function $f : \Omega \subseteq \mathfrak{R}^3 \to \mathfrak{R}$ is integrable on the compact $\Omega$ if and only if for any $\varepsilon > 0$ one can find $\eta = \eta(\varepsilon)$, such that $S_\Delta(f) - s_\Delta(f) < \varepsilon$ for any partition $\Delta$ of norm $v(\Delta) < \eta$.*

**Proof.** It is the same as that from the simple integral. ◘

*Remark.* As always, Darboux's criterion eliminates the "degree of freedom" due to the choice of the intermediate points.

# 4.2. THE CALCULATION OF THE TRIPLE INTEGRAL

As in the case of the double integral, we shall compute the triple integral firstly on simpler domains (parallelepiped, cylinder) and then on simple $Oz$ domains.

## 4.2.1. COMPUTING THE TRIPLE INTEGRAL ON A PARALLELEPIPED

Consider the parallelepiped $\Omega = [a_1, b_1] \times [a_2, b_2] \times [a_3, b_3]$, on which we set up the following partition $\Delta$: we firstly consider partitions on each of the intervals $[a_1, b_1]$, $[a_2, b_2]$, $[a_3, b_3]$, taken on the three axes:

$$a_1 = x_0 < x_1 < x_2 < \ldots x_{i-1} < x_i < \ldots x_n \equiv b_1 \to \delta_1,$$
$$a_2 = y_0 < y_1 < y_2 < \ldots y_{j-1} < y_j < \ldots y_m \equiv b_2 \to \delta_2, \quad (4.2.1)$$
$$a_3 = z_0 < z_1 < z_2 < \ldots z_{k-1} < z_k < \ldots z_p \equiv b_3 \to \delta_3.$$

Then we plot parallel planes to the coordinate planes, passing through the points of division of the three intervals. Thus, the domain $\Omega$ is divided in $mnp$ small parallelepipeds $\Omega_{ijk}$, which form the division $\Delta$, i.e., $\Delta = \left\{ \Omega_{ijk}, i = \overline{1,n},\ j = \overline{1,m},\ k = \overline{1,p} \right\}$ (figure 4.6).

Let us denote

$$\Delta x_i = x_i - x_{i-1},$$
$$\Delta y_j = y_j - y_{j-1}, \quad (4.2.2)$$
$$\Delta z_k = z_k - z_{k-1}.$$

Obviously, the volumes of the parallelepipeds $\Omega_{ijk}$ which form $\Delta$ are

$$\left| \Omega_{ijk} \right| = \Delta \Omega_{ijk} = \Delta x_i \cdot \Delta y_j \cdot \Delta z_k. \quad (4.2.3)$$

***a)*** The case $f(x,y,z) = f_1(x) \cdot f_2(y) \cdot f_3(z).$ $\quad (4.2.4)$

A point $P_{ijk}\left( \alpha_i, \beta_j, \gamma_k \right)$ belongs to $\Omega_{ijk}$ if $x_{i-1} \le \alpha_i \le x_i$, $y_{j-1} \le \beta_i \le y_j$, $z_{k-1} \le \gamma_k \le z_k$. We set up the Riemann sum:

$$\sigma_\Delta\left(f, P_{ijk}\right) = \sum_{i=1}^{n} \sum_{j=1}^{m} \sum_{k=1}^{p} f\left(\alpha_i, \beta_j, \gamma_k\right) \Delta x_i \Delta y_j \Delta z_k =$$
$$= \sum_{i=1}^{n} \sum_{j=1}^{m} \sum_{k=1}^{p} f_1\left(\alpha_i\right) \cdot f_2\left(\beta_j\right) \cdot f_3\left(\gamma_k\right) \Delta x_i \Delta y_j \Delta z. \quad (4.2.5)$$

We notice that, in this case, ***the sums can be separated***:

$$\sigma_\Delta\left(f, P_{ijk}\right)=$$

$$= \underbrace{\sum_{i=1}^{n} f_1\left(\alpha_i\right)\Delta x_i}_{\sigma_{\delta_1}(f_1,\alpha_i)} \cdot \underbrace{\sum_{j=1}^{m} f_2\left(\beta_j\right)\Delta y_j}_{\sigma_{\delta_2}(f_2,\beta_j)} \cdot \underbrace{\sum_{k=1}^{p} f_3\left(\gamma_k\right)\Delta z_k}_{\sigma_{\delta_3}(f_3,\gamma_k)} . \qquad (4.2.6)$$

We see that each one of the three Riemann sums from the above formula tends to a simple integral, i.e.:

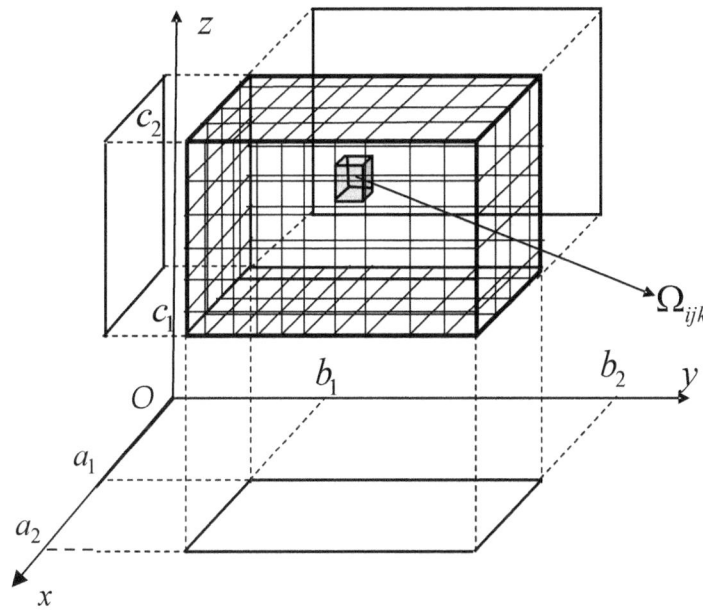

*Figure 4. 6. The partition of the domain for the calculus of the triple integral on a parallelepiped*

$$\sigma_{\delta_1}\left(f_1,\alpha_i\right) \to \int_{a_1}^{b_1} f_1\left(x\right)\mathrm{d}\,x,$$

$$\sigma_{\delta_2}\left(f_2,\beta_j\right) \to \int_{a_2}^{b_2} f_2\left(y\right)\mathrm{d}\,y, \qquad (4.2.7)$$

$$\sigma_{\delta_3}\left(f_3,\gamma_k\right) \to \int_{a_3}^{b_3} f_3\left(z\right)\mathrm{d}\,z .$$

Finally, in this case

$$
\iiint_\Omega f_1(x) \cdot f_2(y) \cdot f_3(z) \, \mathrm{d}x \, \mathrm{d}y \, \mathrm{d}z =
$$
$$
= \int_{a_1}^{b_1} f_1(x) \, \mathrm{d}x \cdot \int_{a_2}^{b_2} f_2(y) \, \mathrm{d}y \cdot \int_{a_3}^{b_3} f_3(z) \, \mathrm{d}z ,
$$

(4.2.8)

therefore, we compute the triple integral as **the product of three simple integrals**.

**b)** If $f = f(x, y, z)$ has not the previous particular form, then we write the Riemann sum in the form:

$$
\sigma_\Delta \left( f, P_{ijk} \right) = \sum_{i=1}^{n} \Delta x_i \sum_{j=1}^{m} \Delta y_j \sum_{k=1}^{p} f\left( \alpha_i, \beta_j, \gamma_k \right) \Delta z_k .
$$

(4.2.9)

Taking limits, we obtain

$$
\iiint_\Omega f(x, y, z) \, \mathrm{d}x \, \mathrm{d}y \, \mathrm{d}z = \int_{a_1}^{b_1} \mathrm{d}x \int_{a_2}^{b_2} \mathrm{d}y \int_{a_3}^{b_3} f(x, y, z) \, \mathrm{d}z .
$$

(4.2.10)

In other words, **to compute a triple integral means to compute successively three simple integrals.**

Therefore, the denomination of **triple integral** is justified not only because it is defined on a three-dimensional domain, but also by its method of computation.

## 4.2.2. THE CALCULUS OF THE TRIPLE INTEGRAL ON A CYLINDER

We now consider the right cylinder $\Omega$, whose bases are situated in the planes $z = a$, $z = b$, both parallel to the plane $xOy$;

their projections on $xOy$ is the plane domain $D$ ( figure 4.7). We set up the following partition $\Delta$:

Firstly, we divide $D$ in subdomains $D_j$, of areas $|D_j|$, and we divide the cylinder $\Omega$ in subcylinders $\Omega_{jk}$, of volumes $|\Omega_{jk}| = \Delta\Omega_{jk}$ and of basis whose projections on $xOy$ are $D_j$, being crossed by planes perpendicular on $Oz$ $\left( j = \overline{1,n}, \ k = \overline{1,m} \right)$. This division is presented in figure 4.7.

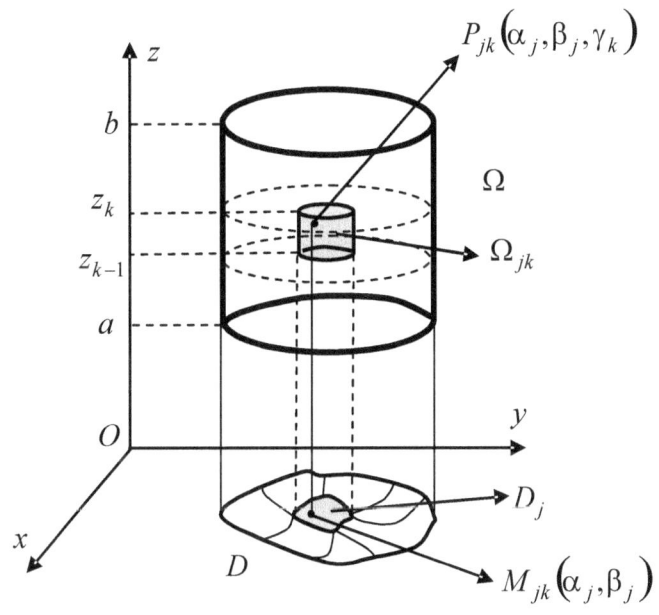

*Figure 4. 7. The partition of the domain for the calculus of the triple integral on a cylinder*

***a)*** Suppose that $f(x, y, z) = f_1(z) \cdot g_1(x, y)$. If $P_{jk} \in \Omega_{jk}$,

then $P_{jk}\left( \alpha_j, \beta_j, \gamma_k \right)$, where $M_j\left( \alpha_j, \beta_j \right) \in D_j$, and $\gamma_k \in \left[ z_{k-1}, z_k \right]$. We set up $\sigma_\Delta \left( f, P_{jk} \right)$ as follows:

$$\sigma_\Delta\left(f,P_{jk}\right)=\sum_{j=1}^{n}\sum_{k=1}^{m}f\left(\alpha_j,\beta_j,\gamma_k\right)\Delta\Omega_{jk}\,. \qquad (4.2.11)$$

But $\Delta\Omega_{jk}=\Delta D_j\cdot\Delta z_k$ is the volume of the right cylinder, in which $\Delta D_j=\left|D_j\right|$ is the notation for the area of $D_j$ and $\Delta z_k=z_k-z_{k-1}$. Therefore, *the Riemann sum separates*:

$$\sigma_\Delta\left(f,P_{jk}\right)=\sum_{j=1}^{n}g_1\left(\alpha_j,\beta_j\right)\Delta D_j\sum_{k=1}^{m}f_1\left(\gamma_k\right)\Delta z_k \qquad (4.2.12)$$

and

$$\sum_{j=1}^{n}g_1\left(\alpha_j,\beta_j\right)\Delta D_j=\sigma_{\Delta_D}\left(g_1,M_j\right),$$

$$\sum_{k=1}^{m}f_1\left(\gamma_k\right)\Delta z_k=\sigma_{\Delta_{[a,b]}}\left(f_1,\gamma_k\right). \qquad (4.2.13)$$

Taking limits, we obtain

$$\iiint_\Omega f_1(z)\cdot g_1(x,y)\,\mathrm{d}x\,\mathrm{d}y\,\mathrm{d}z=\iint_D g_1(x,y)\,\mathrm{d}x\,\mathrm{d}y\int_a^b f_1(z)\,\mathrm{d}z\,. \quad (4.2.14)$$

*b)* Even if $f$ has not the particular form from *a)*, we still can write the Riemann sum as follows

$$\sigma_\Delta\left(f,P_{jk}\right)=\sum_{j=1}^{n}\Delta D_j\sum_{k=1}^{p}f\left(\alpha_j,\beta_j,\gamma_k\right)\Delta z_k, \qquad (4.2.15)$$

which, taking limits, becomes

$$\boxed{\iiint_\Omega f(x,y,z)\,\mathrm{d}x\,\mathrm{d}y\,\mathrm{d}z=\iint_D \mathrm{d}x\,\mathrm{d}y\int_a^b f(x,y,z)\,\mathrm{d}z}\,. \qquad (4.2.16)$$

### 4.2.3. COMPUTING THE TRIPLE INTEGRAL ON SIMPLE $Oz$ DOMAINS

Let $\Omega$ be a simple $Oz$ domain, of boundary formed by the pieces of surface $S_1, S_2$, with the corresponding orientation. We assume that the equations of these surfaces $S_1, S_2$ are

$$\begin{cases} S_1 : z = \varphi_1(x, y), \\ S_2 : z = \varphi_2(x, y), \end{cases} (x, y) \in D. \qquad (4.2.17)$$

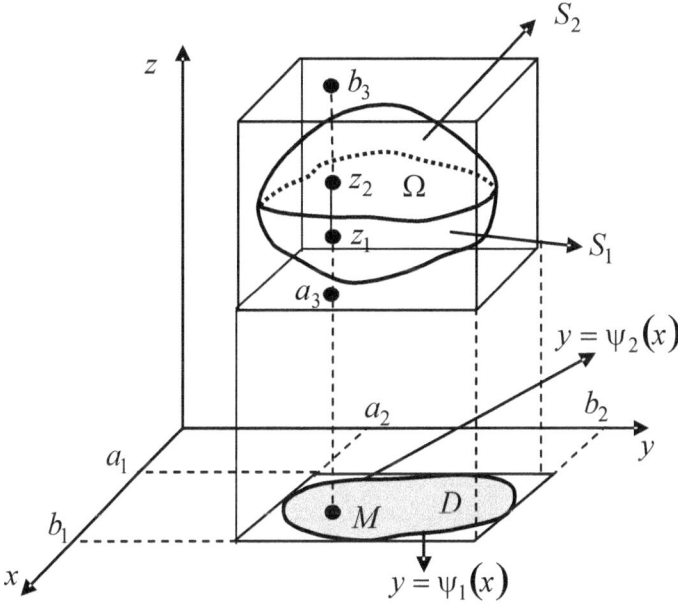

*Figure 4. 8. The simple Oz domain contained in a parallelipiped, projected on the plane xOy*

Let us set up a parallelepiped $V$ containing the domain $\Omega$ (figure 4.8) and define the following auxiliary function

$$f^*(x, y, z) = \begin{cases} f(x, y, z), & \text{on } \Omega, \\ 0, & \text{on } V \setminus \Omega. \end{cases} \qquad (4.2.18)$$

Due to the property of additivity of the triple integral with respect to the domain of integration, we have

$$\iiint_V f^*(x,y,z)\,dx\,dy\,dz =$$

$$= \underbrace{\iiint_\Omega f^*(x,y,z)\,dx\,dy\,dz}_{f(x,y,z)} + \underbrace{\iiint_{V\backslash\Omega} f^*(x,y,z)\,dx\,dy\,dz}_{0}. \quad (4.2.19)$$

Therefore, using formula (4.2.10), it results

$$\iiint_\Omega f(x,y,z)\,dx\,dy\,dz = \iiint_V f^*(x,y,z)\,dx\,dy\,dz \stackrel{(4.2.10)}{=}$$

$$= \int_{a_1}^{b_1}\int_{a_2}^{b_2} dx\,dy \int_{a_3}^{b_3} f^*(x,y,z)\,dz. \quad (4.2.20)$$

But taking into account the spots where $f^*$ cancels, we immediately deduce that:

$$\boxed{\iiint_\Omega f(x,y,z)\,dx\,dy\,dz = \iint_D dx\,dy \int_{\varphi_1(x,y)}^{\varphi_2(x,y)} f(x,y,z)\,dz.} \quad (4.2.21)$$

Moreover, if the domain $D$ is simple $Oy$, then

$$\boxed{\iiint_\Omega f(x,y,z)\,dx\,dy\,dz = \int_a^b dx \int_{\psi_1(x)}^{\psi_2(x)} dy \int_{\varphi_1(x,y)}^{\varphi_2(x,y)} f(x,y,z)\,dz.} \quad (4.2.22)$$

Indeed, if we use the notation $F(x,y) = \displaystyle\int_{\varphi_1(x,y)}^{\varphi_2(x,y)} f(x,y,z)\,dz$,

then from (4.2.21) we obtain

$$\iiint_\Omega f(x,y,z)\,dx\,dy\,dz = \iint_D F(x,y)\,dx\,dy =$$

$$= \int_a^b dx \int_{\psi_1(x)}^{\psi_2(x)} F(x,y)\,dy, \quad (4.2.23)$$

whence we easily get (4.2.22).

*Examples*

1. Compute

$$I = \iiint\limits_{\Omega} \left( x^2 + y^2 + z^2 \right) dx\, dy\, dz , \qquad (4.2.24)$$

where $\Omega$ is a parallelepiped of sides $a, b, c$, its left inferior vertex coinciding with the origin and its edges being parallel to the axes of coordinates (figure 4.9).

**Solution.**

$$I = \int\limits_0^a dx \int\limits_0^b dy \int\limits_0^c \left( x^2 + y^2 + z^2 \right) dz =$$

$$= \int\limits_0^a dx \int\limits_0^b \left( x^2 z + y^2 z + \frac{z^3}{3} \right) \Bigg|_{z=0}^{\,|z=c} dy =$$

$$= \int\limits_0^a dx \int\limits_0^b \left( x^2 c + y^2 c + \frac{c^3}{3} \right) dy = \int\limits_0^a \left( cx^2 y + \frac{y^3}{3} c + \frac{c^3 y}{3} \right) \Bigg|_{y=0}^{\,|y=b} dx = \qquad (4.2.25)$$

$$= \int\limits_0^a \left( x^2 bc + \frac{b^3 c}{3} + \frac{c^3 b}{3} \right) dx = \left( \frac{x^3}{3} bc + \frac{b^3 c}{3} x + \frac{c^3 b}{3} x \right) \Bigg|_{x=0}^{\,|x=a} .$$

Hence

$$\boxed{I = \frac{abc}{3} \left( a^2 + b^2 + c^2 \right) .} \qquad (4.2.26)$$

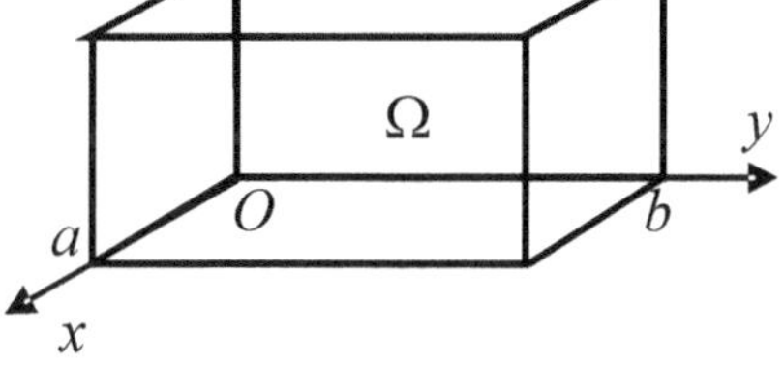

*Figure 4. 9. Parallelepiped of sides a,b,c, with the left inferior vertex at the origin*

2. Compute the volume of the body bounded by the paraboloid $z = \dfrac{x^2 + y^2}{h}$ and by the plane $z = h$.

**Solution.**

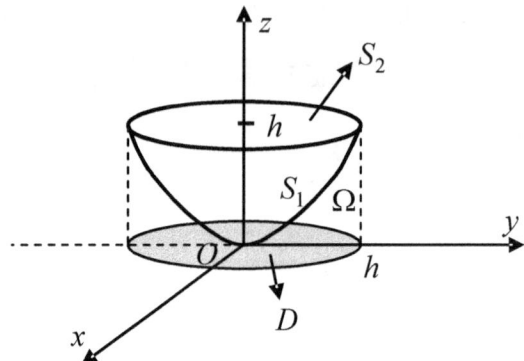

*Figure 4. 10. The paraboloid $z = \dfrac{x^2 + y^2}{h}$ bounded by the plane $z = h$*

The volume of $\Omega$ is

$$|\Omega| = \iiint\limits_{\Omega} dx\,dy\,dz. \tag{4.2.27}$$

The domain is simple $Oz$ and it is bounded by the surfaces:

$$\begin{cases} S_1 : z = \dfrac{x^2 + y^2}{h}, \\ S_2 : z = h. \end{cases} \tag{4.2.28}$$

Therefore

$$\iiint\limits_{\Omega} dx\,dy\,dz = \iint\limits_{D} dx\,dy \int\limits_{\frac{x^2+y^2}{h}}^{h} dz = \iint\limits_{D} z\Big|_{z=\frac{x^2+y^2}{h}}^{z=h} dx\,dy =$$

$$= \iint\limits_{D}\left(-\frac{x^2 + y^2}{h} + h\right)dx\,dy. \tag{4.2.29}$$

$D$ is the projection of $\Omega$ on the plane $xOy$ and it is obviously a circle centered at $O$, of radius $h$. In order to compute the double integral on $D$, we can use polar coordinates:

$$\begin{cases} x = \rho\cos\theta \\ y = \rho\sin\theta \end{cases}, \quad \rho \in [0,h], \theta \in [0,2\pi).$$

In the plane $\rho o\theta$, these coordinates map $D$ into the rectangle $\mathcal{D}$.

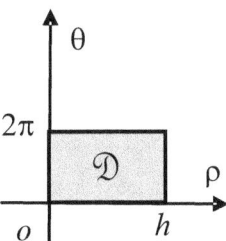

*Figure 4. 11. The domain D mapped into $\mathcal{D}$*

By straightforward computation, we obtain

$$\iint_D \frac{h^2 - \left(x^2 - y^2\right)}{h} dx\, dy = -\frac{1}{h}\iint_{\mathcal{D}}\left(h^2 - \rho^2\right)\rho\, d\rho\, d\theta =$$

$$= \frac{1}{h}\int_0^h \left(h^2 - \rho^2\right)\rho\, d\rho \cdot \int_0^{2\pi} d\theta = \frac{2\pi}{h}\left(h^2\frac{\rho^2}{2} - \frac{\rho^4}{4}\right)\Bigg|_{\rho=0}^{\rho=h} = \quad (4.2.30)$$

$$= \frac{2\pi}{h}\left(\frac{h^4}{2} - \frac{h^4}{4}\right).$$

Hence, the volume of $\Omega$ is

$$\boxed{|\Omega| = \frac{\pi h^3}{2}}. \qquad (4.2.31)$$

3. Compute the volume of the intersection of the cylinders

$$\begin{cases} x^2 + y^2 = a^2 \\ x^2 + z^2 = a^2 \end{cases}$$ situated in the first octant.

**Solution.** The volume of $\Omega$ is

$$|\Omega| = \iiint\limits_{\Omega} dx\,dy\,dz. \qquad (4.2.32)$$

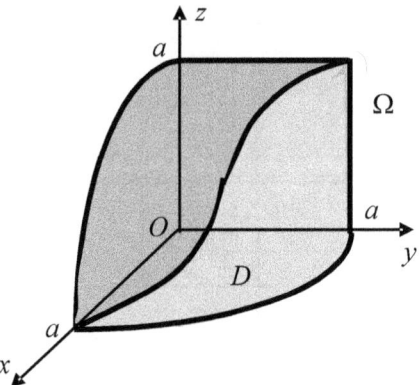

Figure 4. 12. The intersection between the two cylinders:
$$x^2 + y^2 = a^2, x^2 + z^2 = a^2$$

The domain is simple $Oz$ and it is contained between the

surfaces of equations $\begin{cases} S_1 : z = 0, \\ S_2 : z = \sqrt{a^2 - x^2} \end{cases}$. Its projection on $xOy$ is

the quarter of a circle $D$ (figure 4.13).

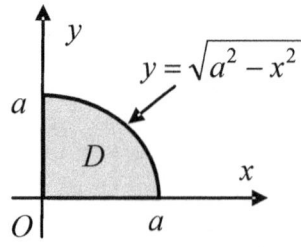

Figure 4. 13. The domain D from the plane xOy

Therefore,

$$|\Omega| = \iint\limits_{D} dx\,dy \int\limits_{0}^{\sqrt{a^2-x^2}} dz = \iint\limits_{D} z\Big|_{z=0}^{z=\sqrt{a^2-x^2}} dx\,dy =$$

$$= \iint\limits_{D} \sqrt{a^2 - x^2}\,dx\,dy .$$

(4.2.33)

We notice that $D$ is simple $Oy$ and it follows that

$$\iint\limits_{D} \sqrt{a^2 - x^2}\,dx\,dy = \int\limits_{0}^{a} dx \int\limits_{0}^{\sqrt{a^2-x^2}} \sqrt{a^2 - x^2}\,dy =$$

$$= \int\limits_{0}^{a} \sqrt{a^2 - x^2} \cdot y \Bigg|_{y=0}^{\left|y=\sqrt{a^2-x^2}\right.} dx = \int\limits_{0}^{a} \left(a^2 - x^2\right) dx =$$

(4.2.34)

$$= a^2 x\Big|_{0}^{a} - \frac{x^3}{3}\Big|_{0}^{a} \Rightarrow \boxed{|\Omega| = \frac{2a^3}{3}} .$$

## 4.3. CHANGES OF VARIABLES IN THE TRIPLE INTEGRAL

Consider the transformation

$$\mathbf{T} : \begin{cases} x = x\left(u, v, w\right) \\ y = y\left(u, v, w\right), \quad \left(u, v, w\right) \in \vartheta \subseteq \Re^{3}, \\ z = z\left(u, v, w\right) \end{cases}$$

(4.3.1)

of the domain $\vartheta$ from $uovw$ to the domain $\Omega$ from $xOyz$.

We assume that $\mathbf{T}$ is nonsingular and that it belongs to $C^{1}\left(\vartheta\right)$.

This means that

$$\frac{D(x,y,z)}{D(u,v,w)} \neq 0, \quad (u,v,w) \in \vartheta. \tag{4.3.2}$$

In this case too, the boundary of $\vartheta$ is mapped into the boundary of $\Omega$. For a better understanding, we also assume that **T** conserves orientation.

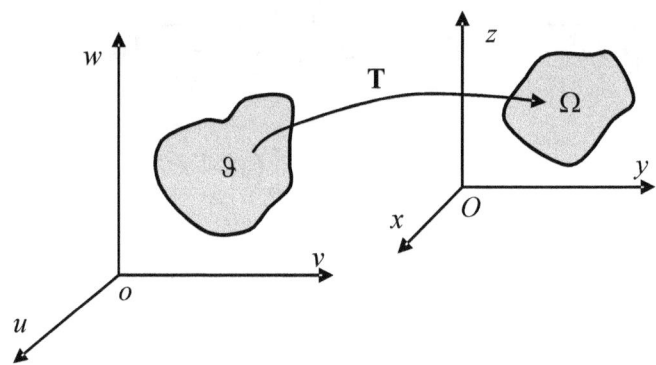

*Figure 4. 14. The transformation **T** of the domain $\vartheta$ into $\Omega$*

Let $f \in C^0(\Omega)$. Then we can prove that

$$\iiint\limits_{\Omega} f(x,y,z)\,dx\,dy\,dz =$$
$$\iiint\limits_{\vartheta} f(x(u,v,w),y(u,v,w),z(u,v,w)) \left|\frac{D(x,y,z)}{D(u,v,w)}\right| du\,dv\,dw. \tag{4.3.3}$$

Indeed, the element of volume of $\Omega$ is transformed as follows:

$$dx\,dy\,dz = \left|\frac{D(x,y,z)}{D(u,v,w)}\right| du\,dv\,dw. \tag{4.3.4}$$

## 4.3.1. CYLINDRICAL COORDINATES

These coordinates (see, e.g., [2,7,10,11]) are very useful for domains with cylindrical symmetry. We recall them.

In this case, the transformation $\mathbf{T}$ is

$$\mathbf{T}:\begin{cases} x = \rho\cos\theta \\ y = \rho\sin\theta, \ \rho \in (0,\infty), \ z \in (-\infty,+\infty), \ \theta \in [0,2\pi). \\ z = z \end{cases} \quad (4.3.5)$$

We compute the Jacobian:

$$\frac{D(x,y,z)}{D(\rho,\theta,z)} = \begin{vmatrix} \dfrac{\partial x}{\partial \rho} & \dfrac{\partial x}{\partial \theta} & \dfrac{\partial x}{\partial z} \\ \dfrac{\partial y}{\partial \rho} & \dfrac{\partial y}{\partial \theta} & \dfrac{\partial y}{\partial z} \\ \dfrac{\partial z}{\partial \rho} & \dfrac{\partial z}{\partial \theta} & \dfrac{\partial z}{\partial z} \end{vmatrix} = \begin{vmatrix} \cos\theta & -\rho\sin\theta & 0 \\ \sin\theta & \rho\cos\theta & 0 \\ 0 & 0 & 1 \end{vmatrix} = \rho. \quad (4.3.6)$$

As $\rho > 0$, it follows that $\left|\dfrac{D(x,y,z)}{D(u,v,w)}\right| = \rho$, therefore in cylindrical coordinates the element of volume is transformed as follows:

$$dx\,dy\,dz = \rho\,d\rho\,d\theta\,dz. \quad (4.3.7)$$

The formula of transformation of the triple integral if we apply cylindrical coordinates is

$$\boxed{\iiint\limits_{\Omega} f(x,y,z)\,dx\,dy\,dz = \iiint\limits_{9} f(\rho\cos\theta,\rho\sin\theta,z)\rho\,d\rho\,d\theta\,dz}. \quad (4.3.8)$$

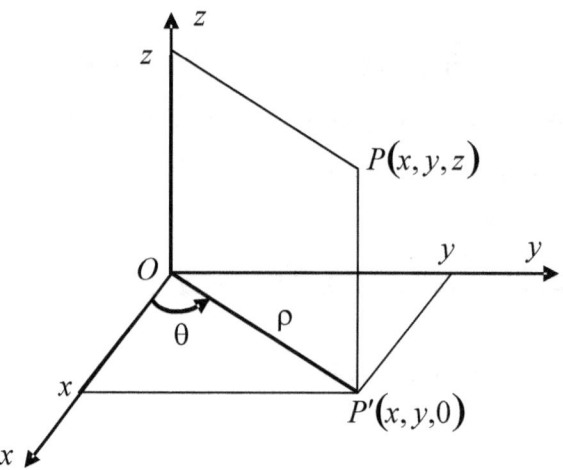

*Figure 4. 15. Cylindrical coordinates*

*Example.* We reconsider the example 2), in which we computed the volume of the body bounded by the parallelepiped $z = \dfrac{x^2 + y^2}{h}$ and by the plane $z = h$.

**Solution.** The volume of $\Omega$ is

$$|\Omega| = \iiint_{\Omega} dx\, dy\, dz.$$ 

$$(4.3.9)$$

The domain has a cylindrical symmetry, therefore we apply cylindrical coordinates. We have

$$|\Omega| = \iiint_{\Omega} \rho\, d\rho\, d\theta\, dz.$$ 

$$(4.3.10)$$

In the plane $\rho o \theta z$, the domain $\vartheta$ is bounded by the surface $z = \dfrac{\rho^2}{h}$ and by the planes $\rho = 0$, $z = h$, $\theta = 0$, $\theta = 2\pi$.

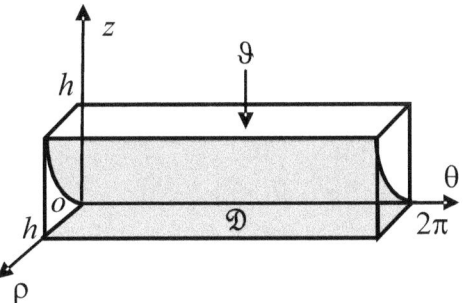

*Figure 4. 16. The domain bounded by the surface* $z = \dfrac{\rho^2}{h}$ *and by the planes*

$$\rho = 0,\ z = h,\ \theta = 0,\ \theta = 2\pi$$

The domain $\vartheta$ is simple $Oz$ and its projection on the plane $z = 0\,(\rho o \theta)$ is the rectangle $\mathcal{D}$. Hence,

$$|\Omega| = \iint_{\mathcal{D}} d\rho\, d\theta \int_{\frac{\rho^2}{h}}^{h} \rho\, dz .$$
(4.3.11)

Computing this, it follows that

$$\iint_{\mathcal{D}} \rho \cdot z\Big|_{z=\frac{\rho^2}{h}}^{z=h} d\rho\, d\theta = \iint_{\mathcal{D}} \left( \rho h - \frac{\rho^3}{h} \right) d\rho\, d\theta,$$
(4.3.12)

such that

$$|\Omega| = \int_{0}^{h} \left( \rho h - \frac{\rho^3}{h} \right) d\rho \cdot \int_{0}^{2\pi} d\theta = 2\pi \left( \frac{\rho^2}{2} h - \frac{\rho^4}{4h} \right)\Bigg|_{\rho=0}^{\rho=h} .$$
(4.3.13)

Finally,

$$\boxed{|\Omega| = \frac{\pi h^3}{2}} .$$
(4.3.14)

## 4.3.2. SPHERICAL COORDINATES

These new coordinates are:

♣ $\rho-$ the distance from the origin to the current point, $\rho \in (0, \infty)$;

♣ $\theta-$ the angle between the $Oz$ axis and the straight line $\overline{OP}$, measured from $Oz$;

In order to have a one-to-one transformation, $\theta \in [0, \pi)$;

♣ $\varphi$ – the angle between the $Ox$ axis and $\overline{OP'}$, where $P'$ is the projection of $P$ on the plane $xOy$, $\varphi \in [0, 2\pi)$.

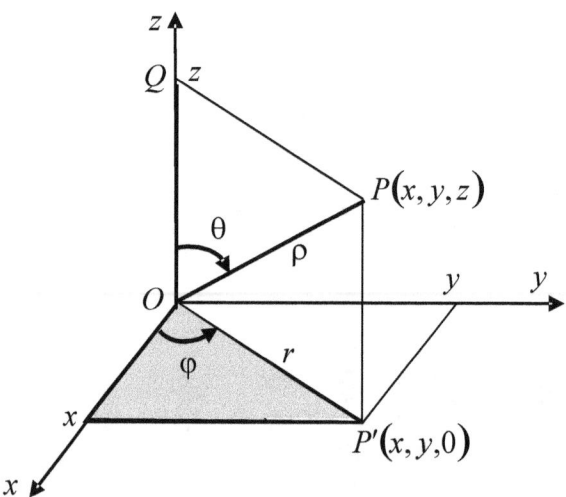

*Figure 4. 17. Spherical coordinates*

Let $r$ be the length of the projection of $\overline{OP}$ on $xOy$, $r = \left| \overline{OP'} \right|$.

Then, in the right-angled triangle from the plane $xOy$ (the shadowed one), we have: $\begin{cases} x = r\cos\varphi \\ y = r\sin\varphi \end{cases}$.

In the triangle $\triangle OPQ$, which has a right angle at $Q$, we notice that $z = \rho\cos\theta$ and $PQ = \rho\sin\theta = r$. Hence, the transformation in spherical coordinates is written as follows:

$$T : \begin{cases} x = \rho\sin\theta\cos\varphi, \\ y = \rho\sin\theta\sin\varphi, \\ z = \rho\cos\theta, \end{cases} \quad \rho \in [0,\infty),\ \theta \in [0,\pi),\ \varphi \in [0,2\pi). \quad (4.3.15)$$

We compute the Jacobian. We have, step by step

$$\frac{D(x,y,z)}{D(\rho,\theta,\varphi)} = \begin{vmatrix} \sin\theta\cos\varphi & \rho\cos\theta\cos\varphi & -\rho\sin\theta\sin\varphi \\ \sin\theta\sin\varphi & \rho\cos\theta\sin\varphi & \rho\sin\theta\cos\varphi \\ \cos\theta & -\rho\sin\theta & 0 \end{vmatrix} =$$

$$= \rho^2[\sin^3\theta\,(\underbrace{\sin^2\varphi + \cos^2\varphi}_{1}) + \qquad\qquad (4.3.16)$$

$$+ \cos^2\theta\sin\theta\,(\underbrace{\cos^2\varphi + \sin^2\varphi}_{1})],$$

and finally

$$\left| \frac{D(x,y,z)}{D(\rho,\theta,\varphi)} \right| = \rho^2\sin\theta > 0, \qquad \theta \in [0,\pi). \qquad (4.3.17)$$

Consequently, the formula of transformation of the triple integral in spherical coordinates is:

$$\iiint_\Omega f(x,y,z)\,dx\,dy\,dz =$$

$$= \iiint_9 f(\rho\sin\theta\cos\varphi, \rho\sin\theta\sin\varphi, \varphi\cos\theta)\rho^2\sin\theta\,d\rho\,d\theta\,d\varphi. \qquad (4.3.18)$$

*Example.* Compute the volume of the upper hemisphere of the sphere of radius $a$, centered at the origin.

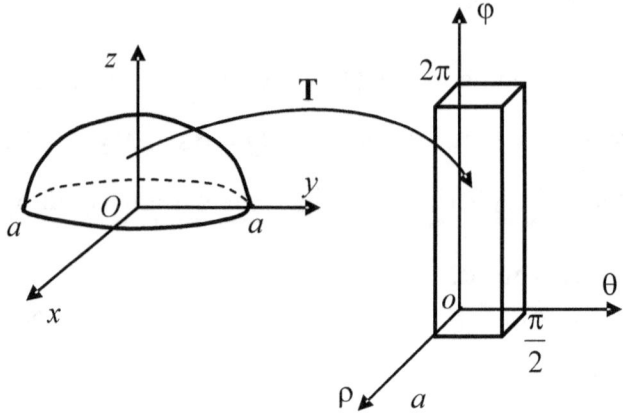

*Figure 4. 18. The transformation of the hemisphere into a parallelepiped*

**Solution.** We compute the volume using the triple integral and spherical coordinates

$$|\Omega| = \iiint_\Omega dx\, dy\, dz = \iiint_9 \rho^2 \sin\theta\, d\rho\, d\theta\, d\varphi. \qquad (4.3.19)$$

The transformed domain is the parallelepiped $[0,a] \times \left[0, \dfrac{\pi}{2}\right] \times [0, 2\pi]$.

We notice that the variables separate within the integrand, therefore the triple integral is the product of three simple integrals:

$$|\Omega| = \int_0^a \rho^2 d\rho \cdot \int_0^{\frac{\pi}{2}} \sin\theta\, d\theta \cdot \int_0^{2\pi} d\varphi = \left.\frac{\rho^3}{3}\right|_0^a \cdot (-\cos\theta)\Big|_0^{\frac{\pi}{2}} \cdot 2\pi$$

$$\Rightarrow \boxed{|\Omega| = \frac{2\pi a^3}{3}}. \qquad (4.3.20)$$

# 4.4. APPLICATIONS OF THE TRIPLE INTEGRAL IN MECHANICS

To make such applications, we must regard tridimensional domains as bodies. With this agreement, we can compute by using the triple integral the following physical quantities:

**1. The volume** of a body $\Omega \subset \mathfrak{R}^3$ is computed as follows

$$|\Omega| = \iiint_{\Omega} dx\,dy\,dz .\qquad (4.4.1)$$

**2. The mass** of a body of density $\gamma(x, y, z) > 0$ is given by

$$m_{\Omega} = \iiint_{\Omega} \gamma(x, y, z)\,dx\,dy\,dz .\qquad (4.4.2)$$

**3. The center of mass** of a body $\Omega \subset \mathfrak{R}^3$ is of coordinates

$$\bar{x} = \frac{\iiint_{\Omega} x\gamma(x, y, z)\,dx\,dy\,dz}{m_{\Omega}},$$

$$\bar{y} = \frac{\iiint_{\Omega} y\gamma(x, y, z)\,dx\,dy\,dz}{m_{\Omega}},\qquad (4.4.3)$$

$$\bar{z} = \frac{\iiint_{\Omega} z\gamma(x, y, z)\,dx\,dy\,dz}{m_{\Omega}}.$$

**4.** The **geometric center of mass** has the coordinates

$$\bar{x} = \frac{\iiint_{\Omega} x\,dx\,dy\,dz}{|\Omega|}, \quad \bar{y} = \frac{\iiint_{\Omega} y\,dx\,dy\,dz}{|\Omega|},$$

$$\bar{z} = \frac{\iiint_{\Omega} z\,dx\,dy\,dz}{|\Omega|}.\qquad (4.4.4)$$

where $|\Omega| = \iiint\limits_{\Omega} dx\,dy\,dz$.

*PARTICULAR CASE:* **the center of mass of a homogeneous body**

If $\gamma(x, y, z) = k$, where $k$ is a positive constant, then we have

$$\overline{x} = \frac{\iiint\limits_{\Omega} xk\,dx\,dy\,dz}{\iiint\limits_{\Omega} k\,dx\,dy\,dz} = \frac{k\iiint\limits_{\Omega} x\,dx\,dy\,dz}{k\iiint\limits_{\Omega} dx\,dy\,dz} = \overline{\overline{x}}, \tag{4.4.5}$$

$$\overline{y} = \overline{\overline{y}}, \quad \overline{z} = \overline{\overline{z}}.$$

Therefore, **the center of mass of a homogeneous body coincides with its geometric center of mass**, i.e.

**THE POSITION OF THE CENTER OF MASS OF A HOMOGENEOUS BODY DEPENDS ONLY ON ITS GEOMETRIC FORM.**

**5.** *Static moments* (with respect to the coordinate planes).

$$M_{xy} = \iiint\limits_{\Omega} z\gamma(x, y, z)\,dx\,dy\,dz,$$

$$M_{xz} = \iiint\limits_{\Omega} y\gamma(x, y, z)\,dx\,dy\,dz, \tag{4.4.6}$$

$$M_{yz} = \iiint\limits_{\Omega} x\gamma(x, y, z)\,dx\,dy\,dz.$$

Obviously, we can compute the center of mass by using the formulas

$$\overline{x} = \frac{M_{yz}}{m_{\Omega}}, \quad \overline{y} = \frac{M_{zx}}{m_{\Omega}}, \quad \overline{z} = \frac{M_{xy}}{m_{\Omega}}. \tag{4.4.7}$$

*Remark.* In fact, formulae (4.4.7) are deduced by using both mechanical concepts and analytical calculus. Formulae (4.4.3) follow consequently.

## 6. *Moments of inertia*

***a)*** with respect to the coordinate planes:

$$I_{xy} = \iiint\limits_{\Omega} z^2 \gamma (x, y, z) dx \, dy \, dz,$$

$$I_{xz} = \iiint\limits_{\Omega} y^2 \gamma (x, y, z) dx \, dy \, dz, \qquad (4.4.8)$$

$$I_{yz} = \iiint\limits_{\Omega} x^2 \gamma (x, y, z) dx \, dy \, dz.$$

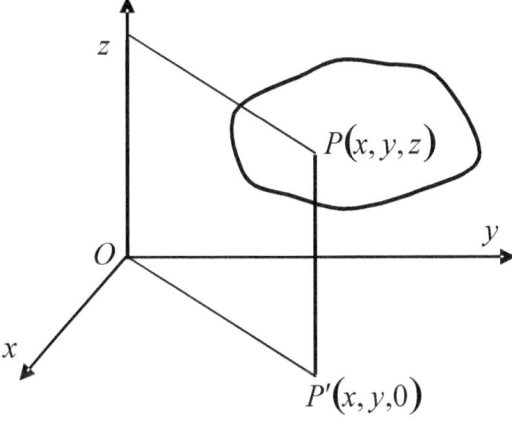

*Figure 4. 19. The distance of the current point P to the plane xOy*

***b)*** with respect to the axes of coordinates:

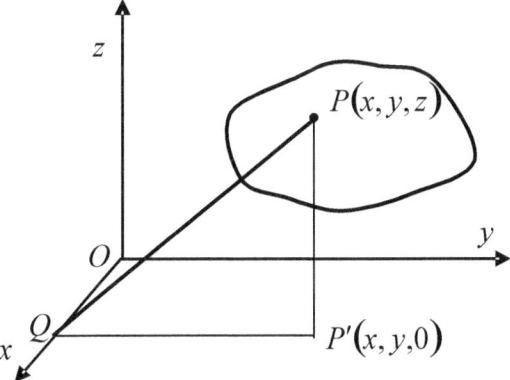

*Figure 4. 20. The distance of the current point P to the Ox axis*

209

According to the theorem of the three perpendiculars, we have

$$\left|\overline{PQ}\right|^2 = \left|\overline{PP'}\right|^2 + \left|\overline{P'Q}\right|^2 = y^2 + z^2. \qquad (4.4.9)$$

It follows that

$$I_x = \iiint_\Omega \left(y^2 + z^2\right) \gamma\left(x, y, z\right) dx\, dy\, dz,$$

$$I_y = \iiint_\Omega \left(x^2 + z^2\right) \gamma\left(x, y, z\right) dx\, dy\, dz, \qquad (4.4.10)$$

$$I_z = \iiint_\Omega \left(x^2 + y^2\right) \gamma\left(x, y, z\right) dx\, dy\, dz.$$

We notice that

$$I_x = I_{xy} + I_{xz},$$
$$I_y = I_{yz} + I_{xy}, \qquad (4.4.11)$$
$$I_z = I_{zx} + I_{zy}.$$

c) with respect to the origin:

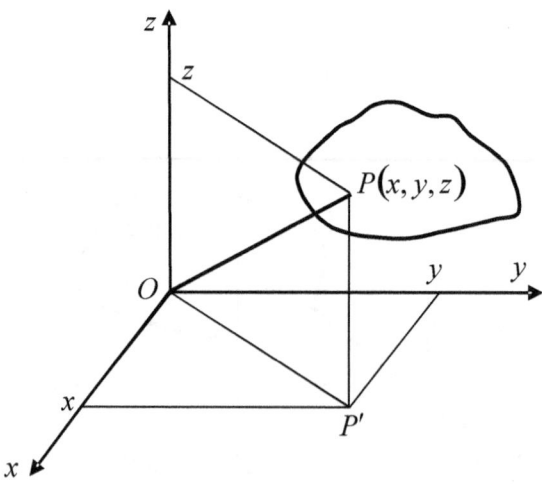

Figure 4. 21. The distance of the current point P to the origin

We notice that $PO^2 = x^2 + y^2 + z^2$, therefore

$$I_O = \iiint_\Omega \left( x^2 + y^2 + z^2 \right) \gamma \left( x, y, z \right) dx\, dy\, dz .$$  (4.4.12)

We also have

$$I_O = I_x + I_{yz} = I_y + I_{xz} = I_z + I_{xy},$$
$$I_O = I_{xy} + I_{yz} + I_{xz}.$$  (4.4.13)

The corresponding **geometric quantities** are obtained by putting $\gamma \left( x, y, z \right) = 1$ in the above formulas.

*Examples*:

1. Compute the geometric moment of inertia with respect to the *Oz* axis of the upper hemisphere of the sphere $x^2 + y^2 + z^2 \leq R^2$.

**Solution.** We apply the spherical coordinates

$$\begin{cases} x = \rho \sin\theta \cos\varphi \\ y = \rho \sin\theta \sin\varphi . \\ z = \rho \cos\theta \end{cases}$$

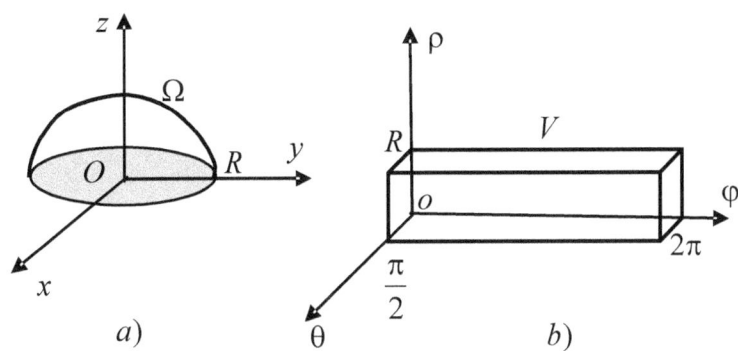

*Figure 4 22. a) The hemisphere $\Omega$; b) Its transformation into the parallelepiped V*

211

We obtain

$$I_z = \iiint\limits_{\Omega} \left(x^2 + y^2\right) dx\, dy\, dz = \iiint\limits_{V} \rho^2 \sin^2\theta \cdot \rho^2 \sin\theta\, d\rho\, d\theta\, d\varphi, \quad (4.4.14)$$

where we made the replacement $\left|\dfrac{D(x,y,z)}{D(\rho,\theta,\varphi)}\right| = \rho^2 \sin\theta$. The

variables separate within the integrand, therefore the triple integral on the parallelepiped $V$ is computed as a product of three simple integrals

$$I_z = \iiint\limits_{V} \rho^4 \sin^3\theta\, d\rho\, d\theta\, d\varphi = \int\limits_0^R \rho^4 d\rho \cdot \int\limits_0^{\frac{\pi}{2}} \sin^3\theta\, d\theta \cdot \int\limits_0^{2\pi} d\varphi =$$

$$= \frac{R^5}{5} \cdot 2\pi \cdot \int\limits_0^{\frac{\pi}{2}} \sin\theta\left(1 - \cos^2\theta\right) d\theta = \quad (4.4.15)$$

$$= \frac{2\pi R^5}{5}\left(-\cos\theta + \frac{\cos^3\theta}{3}\right)\Bigg|_{\theta=0}^{\theta=\frac{\pi}{2}}.$$

It follows that

$$\boxed{I_z = \frac{4\pi R^5}{15}}. \quad (4.4.16)$$

2. Compute the mass of a cylindrical object of density $\gamma(x,y,z) = z\sqrt{x^2 + y^2}$, bounded by the surfaces $x^2 + y^2 = 2x$, $y = 0$, $z = 0$, $z = a$.

**Solution.**

$$m_\Omega = \iiint\limits_{\Omega} z\sqrt{x^2 + y^2}\; dx\, dy\, dz. \quad (4.4.17)$$

We apply cylindrical coordinates

212

$$\begin{cases} x = \rho\cos\theta \\ y = \rho\sin\theta, \quad \rho \in (0,1], z \in [0,a], \theta \in \left[0, \dfrac{\pi}{2}\right]. \\ z = z \end{cases} \qquad (4.4.18)$$

In cylindrical coordinates, the surface $x^2 + y^2 = 2x$ reads $\rho^2 = 2\rho\cos\theta \Rightarrow \rho = 2\cos\theta$.

The Jacobian of the transformation is $\left|\dfrac{D(x,y,z)}{D(\rho,\theta,\varphi)}\right| = \rho$, consequently

$$\iiint_\Omega z\sqrt{x^2 + y^2}\ \mathrm{d}x\,\mathrm{d}y\,\mathrm{d}z = \iiint_\vartheta z\cdot\rho\cdot\rho\ \mathrm{d}\rho\,\mathrm{d}\theta\,\mathrm{d}z, \qquad (4.4.19)$$

where $\vartheta$ is the transformed domain, represented in figure 4. 23.

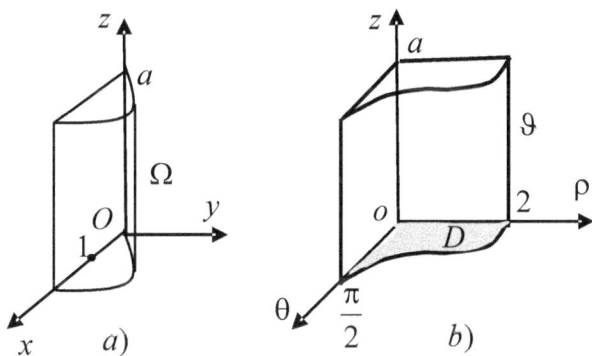

*Figure 4. 23. a) The cylindrical domain $\Omega$; b) Its transformed domain $\vartheta$*

The domain is simple $Oz$ and its projection on the plane $\rho o\theta$ is $D$. It follows that

$$\iiint_\vartheta z\rho^2\mathrm{d}\rho\,\mathrm{d}\theta\,\mathrm{d}z = \iint_D \mathrm{d}\rho\,\mathrm{d}\theta\int_0^a z\rho^2\mathrm{d}z = \frac{a^2}{2}\iint_D \rho^2\mathrm{d}\rho\,\mathrm{d}\theta. \qquad (4.4.20)$$

It remains to compute the double integral. The domain $D$ is simple $o\rho$, therefore

$$m_\Omega = \frac{a^2}{2}\int_0^{\frac{\pi}{2}}\mathrm{d}\theta\int_0^{2\cos\theta}\rho^2\mathrm{d}\rho = \frac{a^2}{2}\int_0^{\frac{\pi}{2}}\left.\frac{\rho^3}{3}\right|_0^{2\cos\theta}\mathrm{d}\theta =$$

$$= \frac{a^2}{2}\cdot\frac{1}{3}\int_0^{\frac{\pi}{2}}8\cos^3\theta\,\mathrm{d}\theta = \frac{a^2}{2}\cdot\frac{8}{3}\int_0^{\frac{\pi}{2}}\cos\theta\left(1-\sin^2\theta\right)\mathrm{d}\theta = \quad(4.4.21)$$

$$= \frac{a^2}{2}\left(\left.\frac{8}{3}\sin\theta\right|_0^{\frac{\pi}{2}} - \left.\frac{8\sin^3\theta}{3}\right|_0^{\frac{\pi}{2}}\right) = \frac{a^2}{2}\left(\frac{8}{3}-\frac{8}{9}\right).$$

It follows that

$$\boxed{m_\Omega = \frac{8}{9}a^2}.\qquad\qquad(4.4.22)$$

3. Compute the mass of the ellipsoid $\dfrac{x^2}{a^2}+\dfrac{y^2}{b^2}+\dfrac{z^2}{c^2}\le 1$, if its

density is $\gamma = \dfrac{x^2}{a^2}+\dfrac{y^2}{b^2}+\dfrac{z^2}{c^2}$.

**Solution.**

$$m_\Omega = \iiint_\Omega \gamma(x,y,z)\,\mathrm{d}x\,\mathrm{d}y\,\mathrm{d}z = \iiint_\Omega\left(\frac{x^2}{a^2}+\frac{y^2}{b^2}+\frac{z^2}{c^2}\right)\mathrm{d}x\,\mathrm{d}y\,\mathrm{d}z\,.\,(4.4.23)$$

We apply **ellipsoidal coordinates** (or **generalized spherical coordinates**):

$$\begin{cases} x = a\rho\cos\varphi\sin\theta \\ y = b\rho\sin\varphi\sin\theta\,, \\ z = c\rho\cos\theta \end{cases} \quad \rho\in[0,1],\ \theta\in[0,\pi),\ \varphi\in[0,2\pi).\quad(4.4.24)$$

The Jacobian is

214

$$\left|\frac{D(x, y, z)}{D(\rho, \theta, \varphi)}\right| =$$

$$= \begin{vmatrix} a\cos\varphi\sin\theta & a\rho\cos\varphi\cos\theta & -a\rho\sin\varphi\sin\theta \\ b\sin\varphi\sin\theta & b\rho\sin\varphi\cos\theta & b\rho\sin\varphi\cos\theta \\ c\cos\theta & -c\rho\sin\theta & 0 \end{vmatrix} = \quad (4.4.25)$$

$$= abc \cdot \rho^2 \sin\theta.$$

The transformed domain is the parallelepiped $\vartheta$.

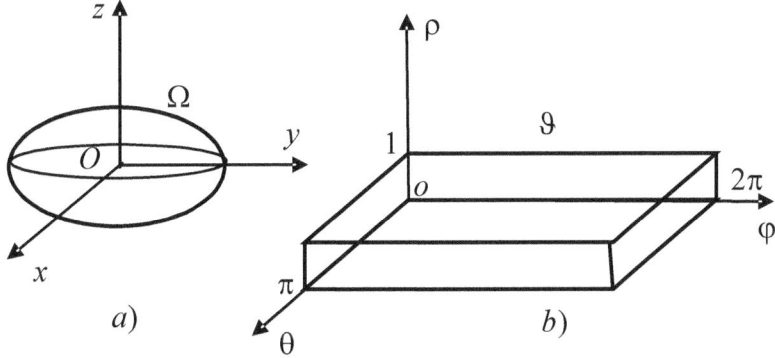

*Figure 4. 24. a) The ellipsoid $\Omega$; b) Its transformed domain, i.e. the parallelepiped $\vartheta$*

$$m_\Omega = \iiint_\vartheta \rho^2 \cdot abc\rho^2 \sin\theta \, d\rho \, d\theta \, d\varphi. \quad (4.4.26)$$

Computing the integrand with respect to the new coordinates, we get

$$\rho^2 \cos^2\varphi\sin^2\theta + \rho^2 \sin^2\varphi\sin^2\theta + \rho^2 \cos^2\theta = \rho^2. \quad (4.4.27)$$

Again, the variables separate within the integrand, hence

$$m_\Omega = abc \int_0^1 \rho^4 d\rho \cdot \int_0^\pi \sin\theta \, d\theta \cdot \int_0^{2\pi} d\varphi =$$

$$= abc \cdot \frac{1}{5} \cdot (-\cos\theta) \Big|_0^\pi \cdot 2\pi \Rightarrow \boxed{m_\Omega = \frac{4\pi abc}{5}}.$$

(4.4.28)

## EXERCISES AND PROBLEMS

1. Compute the following triple integrals on parallelepipeds:

a) $I = \iiint\limits_\Omega (x+y) \, dx \, dy \, dz$,    $\Omega : \begin{cases} 0 \le x \le 2 \\ 0 \le y \le 2 \\ 0 \le z \le 2 \end{cases}$    $A: I = 16$

b) $I = \iiint\limits_\Omega xy^2 z^3 \, dx \, dy \, dz$,    $\Omega : \begin{cases} a_1 \le x \le a_2 \\ b_1 \le y \le b_2 \\ c_1 \le z \le c_2 \end{cases}$

$$A: I = \frac{1}{24} \left( a_2^2 - a_1^2 \right)\left( b_2^3 - b_1^3 \right)\left( c_2^4 - c_1^4 \right)$$

c) $I = \iiint\limits_\Omega \frac{1}{(x+y+z)^2} \, dx \, dy \, dz$,    $\Omega : \begin{cases} 1 \le x \le 3 \\ 0 \le y \le 1 \\ 0 \le z \le 2 \end{cases}$

$$A: I = 4\ln 2 + 6\ln 3 + 5\ln 5 - 8$$

2. Compute the following triple integrals on simple $Oz$ domains:

a) $I = \iiint\limits_\Omega x \, dx \, dy \, dz$, $\Omega \subset \Re^3$ being bounded by the planes $x = 0$, $y = 0$, $z = 0$, $x + y + z = 1$.

$$A: I = \frac{1}{24}$$

b) $I = \iiint\limits_{\Omega} \dfrac{1}{(1+x+y+z)^3}\, dx\, dy\, dz$, $\Omega \subset \mathfrak{R}^3$ being bounded

by the planes $x = 0$, $y = 0$, $z = 0$, $x + y + z = 1$.

$$A: I = \dfrac{1}{2}\ln 2 - \dfrac{5}{16}$$

c) $I = \iiint\limits_{\Omega} \sqrt{x^2 + y^2}\, dx\, dy\, dz$, $\Omega \subset \mathfrak{R}^3$ being bounded by the

surfaces $x + y + z = 2a$, $x^2 + y^2 = a^2$, $z = 0$.

$$A: I = \dfrac{4\pi a^4}{3}$$

3. Compute the volume of the intersection between the paraboloids of equations $x^2 + y^2 = 2 + z$, $x^2 + y^2 = 10 - z$.

$$A: |\Omega| = 36\pi$$

4. Using suitable changes of variables, compute the following integrals:

### I. Using cylindrical coordinates:

a) $I = \iiint\limits_{\Omega} xyz\, dx\, dy\, dz$, $\Omega$ being bounded by the cylinder

of equation $x^2 + y^2 = a^2$ and by the planes $z = 0$, $z = h$.

$$A: I = 0$$

b) The moment of inertia of the body $\Omega$ with respect to $Oz$,

where $\Omega$ is bounded by the cone of equation $x^2 + y^2 = \dfrac{r^2 z^2}{h^2}$ and by

the planes $z = 0$, $z = h$.

$$A: I_z = \dfrac{\pi h r^4}{10}$$

c) The volume of the body bounded by the elliptic paraboloid of equation $x^2 + y^2 = 2z$ and by the plane $z = 2$.

$$A: |\Omega| = 4\pi$$

**II. Using generalized cylindrical coordinates, i.e.,**

$$\begin{cases} x = a\rho\cos\theta \\ y = b\rho\sin\theta : \\ z = z \end{cases}$$

a) Find the static moment with respect to the plane $xOy$ of the body $\Omega$, bounded by the elliptic cylinder of equation $\dfrac{x^2}{a^2} + \dfrac{y^2}{b^2} = 1$ and by the planes $z = 0$, $z = h$.

$$A: M_{xy} = \frac{abh^2\pi}{2}$$

b) Find the moment of inertia of the body $\Omega$ with respect to the plane $yOz$, $\Omega$ being bounded by the cone of equation $\dfrac{x^2}{a^2} + \dfrac{y^2}{b^2} = z^2$ and by the planes $z = 0$, $z = 1$.

$$A: I_{yz} = \frac{a^3 b \pi}{20}$$

c) Compute $I = \iiint\limits_{\Omega} xy \, dx \, dy \, dz$, $\Omega$ being the domain bounded by the cone of equation $\dfrac{x^2}{a^2} + \dfrac{y^2}{b^2} = z^2$ and by the planes $z = 0$, $z = 1$.

$$A: I = 0$$

218

d) Find the coordinates of the center of mass of the body bounded by the cone of equation $\dfrac{x^2}{a^2} + \dfrac{y^2}{b^2} = \dfrac{z^2}{c^2}$ and by the planes $z = 0,\ z = c$.

$$A: \overline{x} = 0, \quad \overline{y} = 0, \quad \overline{z} = \frac{3c}{4}$$

### III. Using spherical coordinates:

a) Compute the mass of the body $\Omega$ bounded by the sphere of equation $x^2 + y^2 + z^2 = z$, if its density is $\gamma(x, y, z) = \sqrt{x^2 + y^2 + z^2}$.

$$A: m_{\Omega} = \frac{\pi}{10}$$

b) Find the volume of the sphere of equation $x^2 + y^2 + z^2 = R^2$.

$$A: |\Omega| = \frac{4R^3\pi}{3}$$

c) Find the moment of inertia of the octant of the sphere $x^2 + y^2 + z^2 = R^2$, $x \geq 0$, $y \geq 0$, $z \geq 0$, with respect to the origin, if its density is $\gamma(x, y, z) = z$.

$$A: I_O = \frac{R^6\pi}{24}$$

### IV. Using generalized spherical coordinates, i.e.,

$$\begin{cases} x = a\rho\sin\theta\cos\varphi \\ y = b\rho\sin\theta\sin\varphi : \\ z = c\rho\cos\theta \end{cases}$$

a) Find the volume of the ellipsoid of equation $\dfrac{x^2}{a^2} + \dfrac{y^2}{b^2} + \dfrac{z^2}{c^2} = 1$.

$$A: |\Omega| = \frac{4abc\pi}{3}$$

b) Find the volume of the body bounded by the ellipsoid of equation $\dfrac{x^2}{a^2} + \dfrac{y^2}{b^2} + \dfrac{z^2}{c^2} = 1$ and by the elliptic cone of equation $\dfrac{x^2}{a^2} + \dfrac{y^2}{b^2} = \dfrac{z^2}{c^2}$.

$$A: |\Omega| = \frac{2abc\pi}{3}\left(1 - \frac{\sqrt{2}}{2}\right)$$

*Hint*: $\vartheta = \left\{ (\rho, \theta, \varphi), 0 \le \rho \le 1, 0 \le \theta \le \dfrac{\pi}{4}, 0 \le \varphi \le 2\pi \right\}$

c) The moment of inertia of the ellipsoid of equation $\dfrac{x^2}{a^2} + \dfrac{y^2}{b^2} + \dfrac{z^2}{c^2} = 1$, with respect to the plane $xOy$.

$$A: I_{xy} = \frac{4\pi abc^3}{15}$$

# Chapter 5

## SURFACE INTEGRALS. INTEGRAL FORMULAS

### 5.1. THE DEFINITION OF THE SURFACE ELEMENT

Consider the transformation $\mathbf{T}: \begin{cases} x = x(u,v) \\ y = y(u,v), \\ z = z(u,v) \end{cases} u,v \in D$. Let

us set up a network of rightlines $u = \text{const}, v = \text{const}$, in the plane $uov$, which is mapped by $\mathbf{T}$ into a network of curves on the surface $S$ (figure 5.1).

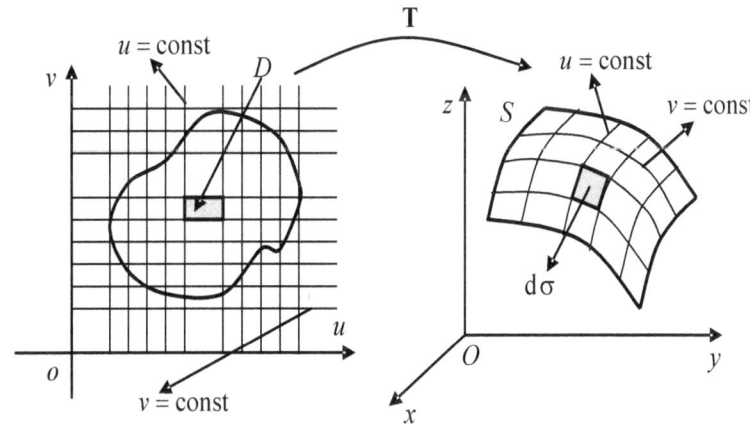

*Figure 5. 1. The transformation of the network $u = \text{const}, v = \text{const}$ into a network of curves on the surface S*

Let us compute **the element of surface area**, denoted by $d\sigma$; in order to do that, we shall, firstly, compute **the element of distance** along $S$, $ds$.

More precisely, the element of distance $ds$ represents the distance between two points $M(x, y, z), M'(x + dx, y + dy, z + dz)$ on $S$, close to each other ( figure 5.2). The bracket $(u, v)$ refers to the calculus of the coordinates $x, y, z$ at the point $(u, v)$.

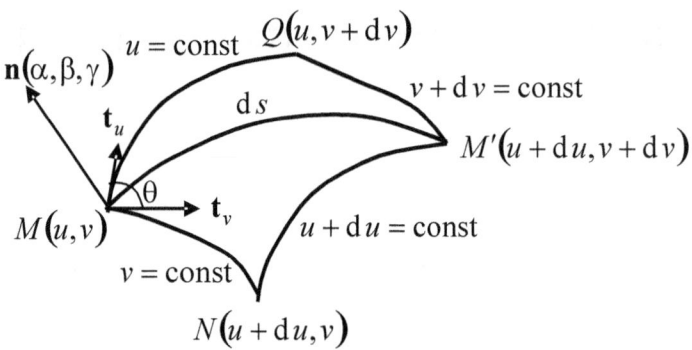

*Figure 5. 2. The element of distance along a surface*

We have

$$\left|\overline{MM'}\right|^2 = ds^2 = dx^2 + dy^2 + dz^2. \tag{5.1.1}$$

But

$$dx = \frac{\partial x}{\partial u}du + \frac{\partial x}{\partial v}dv, \quad dy = \frac{\partial y}{\partial u}du + \frac{\partial y}{\partial v}dv,$$

$$dz = \frac{\partial z}{\partial u}du + \frac{\partial z}{\partial v}dv. \tag{5.1.2}$$

Squaring and adding the relations (5.1.2), we obtain

$$ds^2 = \left[ \left( \frac{\partial x}{\partial u} \right)^2 + \left( \frac{\partial y}{\partial u} \right)^2 + \left( \frac{\partial z}{\partial u} \right)^2 \right] du^2 +$$

$$+ 2 \left[ \frac{\partial x}{\partial u} \cdot \frac{\partial x}{\partial v} + \frac{\partial y}{\partial u} \cdot \frac{\partial y}{\partial v} + \frac{\partial z}{\partial u} \cdot \frac{\partial z}{\partial v} \right] du \, dv + \tag{5.1.3}$$

$$+ \left[ \left( \frac{\partial x}{\partial v} \right)^2 + \left( \frac{\partial y}{\partial v} \right)^2 + \left( \frac{\partial z}{\partial v} \right)^2 \right] dv^2 .$$

Let us use the following notations for the above square brackets (this is a notation commonly used in differential geometry):

$$\begin{cases} E = \left( \dfrac{\partial x}{\partial u} \right)^2 + \left( \dfrac{\partial y}{\partial u} \right)^2 + \left( \dfrac{\partial z}{\partial u} \right)^2 , \\[3mm] F = \dfrac{\partial x}{\partial u} \cdot \dfrac{\partial x}{\partial v} + \dfrac{\partial y}{\partial u} \cdot \dfrac{\partial y}{\partial v} + \dfrac{\partial z}{\partial u} \cdot \dfrac{\partial z}{\partial v} , \\[3mm] G = \left( \dfrac{\partial x}{\partial v} \right)^2 + \left( \dfrac{\partial y}{\partial v} \right)^2 + \left( \dfrac{\partial z}{\partial v} \right)^2 . \end{cases} \tag{5.1.4}$$

It follows that

$$ds^2 = E \, du^2 + 2F \, du \, dv + G \, dv^2 , \tag{5.1.5}$$

which is also called **the first fundamental form** of the surface $S$.

Let us now compute **the area of the elementary parallelogram** $MNM'Q$, i.e. $d\sigma$. From the elementary geometry, it is known that

$$d\sigma = \left| \overline{MN} \right| \cdot \left| \overline{MQ} \right| \cdot \sin\theta . \tag{5.1.6}$$

We apply formula (5.1.3). Along $\overline{MN}$, $v = \text{const}$, so that $dv = 0$. It follows that

$$\overline{MN} = \sqrt{E} \, du . \tag{5.1.7}$$

Along $\overline{MQ}$, $u = \text{const}$, therefore $du = 0$. It follows that

$$\overline{MQ} = \sqrt{G}dv. \qquad (5.1.8)$$

We notice that

a) the direction of the tangent $\mathbf{t}_v$ to the curve $v = \mathrm{const}$ has

the components $\left( \dfrac{\partial x}{\partial u}, \dfrac{\partial y}{\partial u}, \dfrac{\partial z}{\partial u} \right)$, and

b) the direction of the tangent $\mathbf{t}_u$ to the curve $u = \mathrm{const}$ has

the components $\left( \dfrac{\partial x}{\partial v}, \dfrac{\partial y}{\partial v}, \dfrac{\partial z}{\partial v} \right)$.

The corresponding unit vectors (the versors) are obtained by dividing these vectors by their corresponding length. These lengths are:

$$\sqrt{\left( \frac{\partial x}{\partial u} \right)^2 + \left( \frac{\partial y}{\partial u} \right)^2 + \left( \frac{\partial z}{\partial u} \right)^2} = \sqrt{E},$$

$$\sqrt{\left( \frac{\partial x}{\partial v} \right)^2 + \left( \frac{\partial y}{\partial v} \right)^2 + \left( \frac{\partial z}{\partial v} \right)^2} = \sqrt{G}. \qquad (5.1.9)$$

The corresponding direction cosines are expressed as follows:

$$\text{for } \mathbf{t}_v : \left( \frac{\pm\dfrac{\partial x}{\partial u}}{\sqrt{E}}, \frac{\pm\dfrac{\partial y}{\partial u}}{\sqrt{E}}, \frac{\pm\dfrac{\partial z}{\partial u}}{\sqrt{E}} \right), \quad \text{for } \mathbf{t}_u : \left( \frac{\pm\dfrac{\partial x}{\partial v}}{\sqrt{G}}, \frac{\pm\dfrac{\partial y}{\partial v}}{\sqrt{G}}, \frac{\pm\dfrac{\partial z}{\partial v}}{\sqrt{G}} \right). \quad (5.1.10)$$

The cosine of the angle $\theta$ formed by the two directions is

$$\cos\theta = \pm \frac{\dfrac{\partial x}{\partial u} \cdot \dfrac{\partial x}{\partial v} + \dfrac{\partial y}{\partial u} \cdot \dfrac{\partial y}{\partial v} + \dfrac{\partial z}{\partial u} \cdot \dfrac{\partial z}{\partial v}}{\sqrt{EG}} = \pm \frac{F}{\sqrt{EG}}. \qquad (5.1.11)$$

Hence

$$\sin\theta = \pm\sqrt{1 - \cos^2\theta} = \pm\sqrt{1 - \frac{F^2}{EG}} = \pm\frac{\sqrt{EG - F^2}}{\sqrt{EG}}. \qquad (5.1.12)$$

Getting back to the element of area $d\sigma$, we infer that

$$d\sigma = \sqrt{E}\cdot\sqrt{G}\cdot\frac{\sqrt{EG - F^2}}{\sqrt{EG}}du\,dv \Rightarrow d\sigma = \sqrt{EG - F^2}\,du\,dv. \quad (5.1.13)$$

**Definition 5.1.** The area of $S$ is computed by the formula

$$\boxed{|S| = \iint_D \sqrt{EG - F^2}\,du\,dv}. \qquad (5.1.14)$$

It remains to prove that:

1. $\sqrt{EG - F^2} > 0$ and that

2. the definition 5.1 remains unchanged, no matter what change of variables $(u, v)$ we use.

**\* Proof.**

1. Obviously, $\mathbf{n} \perp \mathbf{t}_u$, where $\mathbf{n}(\alpha, \beta, \gamma)$ is the unit vector of the outward normal to $S$. It follows that $\langle \mathbf{n}, \mathbf{t}_u \rangle = 0$, i.e.

$$\alpha\cdot\frac{\partial x}{\partial v} + \beta\cdot\frac{\partial y}{\partial v} + \gamma\cdot\frac{\partial z}{\partial v} = 0. \qquad (5.1.15)$$

But we also have $\langle \mathbf{n}, \mathbf{t}_v \rangle = 0$, because $\mathbf{n} \perp \mathbf{t}_v$. We deduce

$$\alpha\cdot\frac{\partial x}{\partial u} + \beta\cdot\frac{\partial y}{\partial u} + \gamma\cdot\frac{\partial z}{\partial u} = 0. \qquad (5.1.16)$$

The equalities (5.1.15) and (5.1.16) are equivalent to

$$\frac{\alpha}{A} = \frac{\beta}{B} = \frac{\gamma}{C}, \qquad (5.1.17)$$

where

$$A = \frac{D(y,z)}{D(u,v)}, \quad B = \frac{D(z,x)}{D(u,v)}, \quad C = \frac{D(x,y)}{D(u,v)}, \qquad (5.1.18)$$

are the specified Jacobians. After an elementary calculation, we get

$$A^2 + B^2 + C^2 = EG - F^2, \qquad (5.1.19)$$

hence the point 1 is proven.

2. Take two different representations of the surface $S$, in two different systems of coordinates:

$$S : \begin{cases} x = f(u,v) \\ y = g(u,v), \quad (u,v) \in D, \\ z = h(u,v) \end{cases}$$

$$S : \begin{cases} x = f'(u',v') \\ y = g'(u',v'), \quad (u',v') \in D', \\ z = h'(u',v') \end{cases} \qquad (5.1.20)$$

where $f'(u',v') = f(u(u',v'), v(u',v'))$ and the same holds true for $g'$, $h'$. The change of variables between the coordinates $(u,v)$ and $(u',v')$ is

$$T : \begin{cases} u' = u'(u,v) \\ v' = v'(u,v) \end{cases}, \quad (u,v) \in D, \qquad \frac{D(u',v')}{D(u,v)} \neq 0 \text{ in } D. \quad (5.1.21)$$

According to the definition 5.1, we have, on the one hand

$$|S| = \iint_D \sqrt{EG - F^2} \, du \, dv, \qquad (5.1.22)$$

and, on the other hand, the same area $|S|$ is expressed by

$$|S| = \iint_{D'} \sqrt{E'G' - F'^2} \, du' \, dv', \qquad (5.1.23)$$

where $E', F', G'$ correspond to $E, F, G$ respectively, through the representation of $S$ by using the variables $u', v'$.

### *Do these two expressions coincide?*

To be sure that they really coincide, we start from the relations

$$EG - F^2 = A^2 + B^2 + C^2,$$
$$E'G' - F'^2 = A'^2 + B'^2 + C'^2. \tag{5.1.24}$$

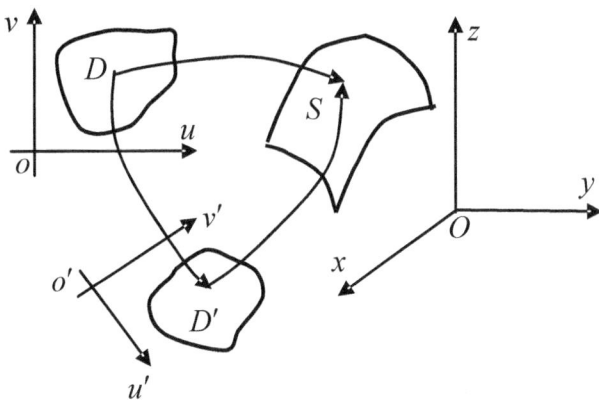

*Figure 5. 3. The change of coordinates* $\mathbf{T} : (u', v') \to (u, v)$

Due to the properties of the Jacobian, we have

$$A' = \frac{D(g,h)}{D(u',v')} = \frac{D(g,h)}{D(u,v)} \cdot \frac{D(u,v)}{D(u',v')} = K\frac{D(g,h)}{D(u,v)} = KA, \tag{5.1.25}$$

where we denoted the Jacobian of the transformation of the variables $(u, v), (u', v')$ by $K$.

Obviously, $K \neq 0$. Analogously, $B' = KB$, $C' = KC$.

Let us apply the change of variables $\mathbf{T}$ to the double integral

$$|S| = \iint\limits_{D'} \sqrt{E'G' - F'^2}\, du'\, dv'. \tag{5.1.26}$$

227

We have

$$\iint_{D'} \sqrt{E'G' - F'^2}\, du'\, dv' = \iint_{D'} \sqrt{A'^2 + B'^2 + C'^2}\, du'\, dv' =$$

$$= \iint_{D} |K|\sqrt{A^2 + B^2 + C^2} \cdot \left| \frac{D(u',v')}{D(u,v)} \right| du\, dv. \tag{5.1.27}$$

But by virtue of the properties of the Jacobian,

$$\underbrace{\frac{D(u,v)}{D(u',v')}}_{K} \cdot \frac{D(u',v')}{D(u,v)} = 1, \tag{5.1.28}$$

hence,

$$\frac{D(u',v')}{D(u,v)} = K^{-1}. \tag{5.1.29}$$

Getting back to the double integral, we obtain

$$\iint_{D'} \sqrt{E'G' - F'^2}\, du'\, dv' = \iint_{D} \underbrace{\sqrt{A^2 + B^2 + C^2}}_{\sqrt{EG-F^2}} \cdot |K| \cdot \frac{1}{|K|} du\, dv, \tag{5.1.30}$$

and thus

$$\iint_{D'} \sqrt{E'G' - F'^2}\, du'\, dv' = \iint_{D} \sqrt{EG - F^2}\, du\, dv. \tag{5.1.31}$$

It follows that ***the definition 5.1*** of the area of a portion of surface ***is correct***, because it does not depend on the parametrization of the surface $S$.


## 5.2. THE SURFACE INTEGRAL OF THE FIRST KIND

Let us notice that:

* ♣ If we start from the simple integral defined on an interval, we bring forward two other types of integrals, i.e. the curvilinear

integral of the first and of the second kind, using the deformation of the domain of definition.

And in the case of double integrals,

♣ If we start from the double integral, we bring forward two types of surface integrals: of the first and of the second kind, also by using the deformation of the domain of definition.

*Figure 5. 4. The deformation of the interval, leading from the simple integral to the curvilinear one*

Intuitively,

♣ the first kind curvilinear integral corresponds to ***mass quantities*** and

♣ the second kind curvilinear integral corresponds to ***dynamic quantities***.

Analogously,

♣ the ***first kind surface integral*** corresponds to ***mass quantities*** and

♣ the ***second kind surface integral*** corresponds to ***dynamic quantities***.

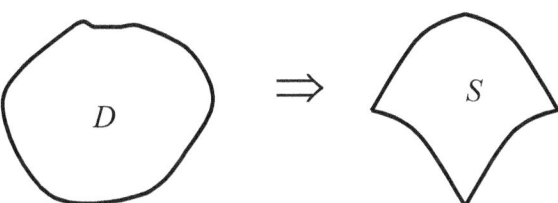

*Figure 5. 5. The deformation of the plane domain, leading from the double integral to the surface integral*

Let $S$ be a portion of surface of area $|S|$. Let us consider the partition $\Delta = \{S_1, S_2, \ldots, S_n\}$ on $S$, having the property that $S = \bigcup\limits_{i=1}^{n} S_i$ and that $S_i$, $S_j$ have, at most, boundary points in common, for $i \neq j$. We use the notation $\Delta\sigma_i = |S_i|$.

**The norm** (mesh size) of $\Delta$ is

$$v(\Delta) = \max\left\{\operatorname{diam} S_i, i = \overline{1, n}\right\}, \tag{5.2.1}$$

where

$$\operatorname{diam} S_i = \max\left\{\left|\overline{PP'}\right|, P, P' \in fr S_i\right\}. \tag{5.2.2}$$

**Warning!** $\overset{\frown}{PP'}$ **must be entirely contained in $S$ !**

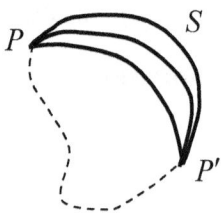

*Figure 5. 6. Measuring the diameter of S*

For example, in figure 5.6., the dotted arc $\overset{\frown}{PP'}$ **does not belong to $S$**, therefore it is not taken into consideration when we measure the diameter of $S$!

Let $f : S \to \Re$. On each $S_i$ we choose $P_i \in S_i, P_i\left(\alpha_i, \beta_i, \gamma_i\right)$. We set up **the Riemann sum**:

$$\sigma_\Delta\left(f, P_i\right) = \sum\limits_{i=1}^{n} f\left(\alpha_i, \beta_i, \gamma_i\right)\Delta\sigma_i. \tag{5.2.3}$$

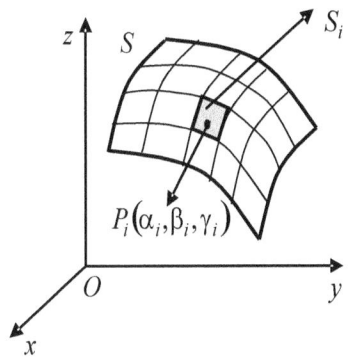

*Figure 5. 7. The partition $\Delta$ on $S$*

**Definition 5. 2.** If there exists a number $I \in \mathfrak{R}$, such that for any $\varepsilon > 0$ one can find $\eta = \eta(\varepsilon)$ with the property that $|\sigma_\Delta(f, P_i) - I| < \varepsilon$, for any partition $\Delta$ of norm $\nu(\Delta) < \eta$ and for any choice of the points $P_i$, then we say that *f is **integrable*** on $S$ and

$$I = \iint_S f(x, y, z) \mathrm{d}\sigma \qquad (5.2.4)$$

is the ***surface integral of the first kind*** of $f$ on $S$.

*Remark.* We notice that the logical structure of this definition is the same as for all the integrals defined in this book.

If $S$ is represented on $D \subset \mathfrak{R}^2$ by the equations

$$S: \begin{cases} x = f(u, v) \\ y = g(u, v), \quad (u, v) \in D, \\ z = h(u, v) \end{cases} \qquad (5.2.5)$$

then the surface integral is computed by using the double integral on $D$:

$$\iint_D f(x, y, z)\,d\sigma =$$

$$= \iint_D f(x(u, v), y(u, v), z(u, v))\sqrt{EG - F^2}\,du\,dv. \tag{5.2.6}$$

**SIGNIFICANT PARTICULAR CASE:**

$S$ is expressed as $z = z(x, y), \ \in D \subset xOy$.

Then $x = u, \ y = v$ and $D$ is the projection of $S$ on $xOy$ (figure 5. 8).

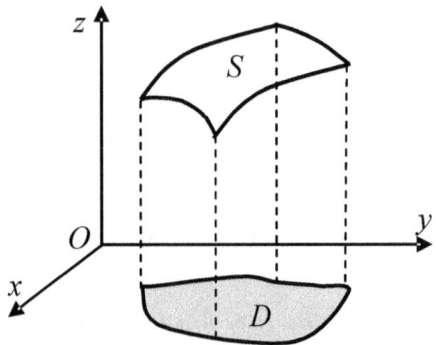

*Figure 5. 8. The projection of $S$ on the plane $xOy$*

We compute

$$E = \left(\frac{\partial x}{\partial x}\right)^2 + \left(\frac{\partial y}{\partial x}\right)^2 + \left(\frac{\partial z}{\partial x}\right)^2 = 1 + \left(\frac{\partial z}{\partial x}\right)^2,$$

$$G = \left(\frac{\partial x}{\partial y}\right)^2 + \left(\frac{\partial y}{\partial y}\right)^2 + \left(\frac{\partial z}{\partial y}\right)^2 = 1 + \left(\frac{\partial z}{\partial y}\right)^2, \tag{5.2.7}$$

$$F = \frac{\partial x}{\partial x} \cdot \frac{\partial x}{\partial y} + \frac{\partial y}{\partial x} \cdot \frac{\partial y}{\partial y} + \frac{\partial z}{\partial x} \cdot \frac{\partial z}{\partial y} = \frac{\partial z}{\partial x} \cdot \frac{\partial z}{\partial y}.$$

Therefore,

$$EG - F^2 = \left[1 + \left(\frac{\partial z}{\partial x}\right)^2\right]\left[1 + \left(\frac{\partial z}{\partial y}\right)^2\right] - \left(\frac{\partial z}{\partial x}\right)^2 \cdot \left(\frac{\partial z}{\partial y}\right)^2 =$$

$$= 1 + \left(\frac{\partial z}{\partial x}\right)^2 + \left(\frac{\partial z}{\partial y}\right)^2 + \left(\frac{\partial z}{\partial x}\right)^2 \cdot \left(\frac{\partial z}{\partial y}\right)^2 - \left(\frac{\partial z}{\partial x}\right)^2 \cdot \left(\frac{\partial z}{\partial y}\right)^2 ,$$

(5.2.8)

i.e.

$$EG - F^2 = 1 + p^2 + q^2 ,$$ 

(5.2.9)

where we used Monge's notations:

$$\frac{\partial z}{\partial x} = p, \quad \frac{\partial z}{\partial y} = q .$$ 

(5.2.10)

Formula (5.2.6) becomes:

$$\boxed{\iint_S f(x, y, z)\,d\sigma = \iint_D f\left(x, y, z(x, y)\right)\sqrt{1 + p^2 + q^2}\ dx\,dy.}$$ 

(5.2.11)

### 5.2.1. PROPERTIES OF THE FIRST KIND SURFACE INTEGRAL

These properties are the same as in the case of the double integral, if we take into account the corresponding specifications.

Let us notice that for $f \equiv 1$, the Riemann sum becomes

$$\sigma_\Delta (1, P_i) = \sum_{i=1}^{n} \Delta\sigma_i = |S|,$$ 

(5.2.12)

for any partition $\Delta$ of $S$ and any choice of the points $P_i$.

Therefore, it follows that the area of $S$ is computed using the first kind surface integral

$$|S| = \iint_S d\sigma .$$ 

(5.2.13)

If the surface is of equations

$$S : \begin{cases} x = f(u,v) \\ y = g(u,v), \quad (u,v) \in D, \\ z = h(u,v) \end{cases} \quad (5.2.14)$$

then

$$\iint_S \mathrm{d}\sigma = \iint_D \sqrt{EG - F^2}\, \mathrm{d}u\, \mathrm{d}v. \quad (5.2.15)$$

## 5.2.2. APPLICATIONS OF THE FIRST KIND SURFACE INTEGRAL IN MECHANICS

To apply surface integral in mechanics, one must regard the portions of surface as **shells**. Shells are, in fact, curved plates; more precisely, they are three-dimensional structures with only two significant dimensions, one of them being negligible.

*1.* If $\gamma = \gamma(x, y, z) > 0$, where $\gamma : S \to \Re$ is regarded as the superficial density of the shell $S$, then

$$m_S = \iint_S \gamma(x, y, z)\, \mathrm{d}\sigma \quad (5.2.16)$$

is *the mass of the shell S*.

*2. The center of mass* of $S$ has the coordinates

$$\overline{x} = \frac{\iint_S x\gamma(x, y, z)\, \mathrm{d}\sigma}{m_S}, \quad \overline{y} = \frac{\iint_S y\gamma(x, y, z)\, \mathrm{d}\sigma}{m_S},$$

$$\overline{z} = \frac{\iint_S z\gamma(x, y, z)\, \mathrm{d}\sigma}{m_S}. \quad (5.2.17)$$

### 3. *The geometric center of mass* of S has the coordinates

$$\overline{x} = \frac{\displaystyle\iint_S x\,d\sigma}{|S|}, \qquad \overline{y} = \frac{\displaystyle\iint_S y\,d\sigma}{|S|}, \qquad \overline{z} = \frac{\displaystyle\iint_S z\,d\sigma}{|S|}, \qquad (5.2.18)$$

where $|S|$ is the area of the portion of surface S.

### SIGNIFICANT PARTICULAR CASE:

If the shell S is homogeneous, therefore if $\gamma = \mathrm{const}$, then, as in the case of the double or triple integral, it follows that $\overline{\overline{x}} = \overline{x}, \quad \overline{\overline{y}} = \overline{y}, \quad \overline{\overline{z}} = \overline{z}$, hence

***THE GEOMETRIC CENTER OF MASS OF A HOMOGENEOUS SHELL COINCIDES WITH ITS CENTER OF MASS.***

### 4. *Moments of inertia*

*a)* with respect to the planes of coordinates:

$$I_{xy} = \iint_S z^2\gamma\,(x, y, z)\,d\sigma, \quad I_{yz} = \iint_S x^2\gamma\,(x, y, z)\,d\sigma,$$

$$I_{zx} = \iint_S y^2\gamma\,(x, y, z)\,d\sigma. \qquad (5.2.19)$$

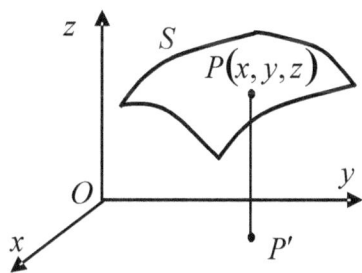

*Figure 5. 9. The distance from the current point to the plane xOy*

*b)* with respect to the axes of coordinates. From the figure 5.10, we notice that

$$PM^2 = MP'^2 + PP'^2 = y^2 + z^2,$$ (5.2.20)

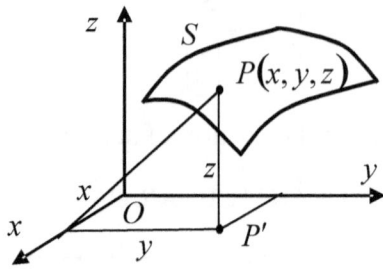

*Figure 5. 10. The distance from the current point P to the Ox axis*

therefore,

$$I_x = \iint_S \left( y^2 + z^2 \right) \gamma (x, y, z) \mathrm{d}\sigma = I_{xy} + I_{xz},$$

$$I_y = \iint_S \left( x^2 + z^2 \right) \gamma (x, y, z) \mathrm{d}\sigma = I_{xy} + I_{yz},$$ (5.2.21)

and

$$I_z = \iint_S \left( x^2 + y^2 \right) \gamma (x, y, z) \mathrm{d}\sigma = I_{xz} + I_{yz}.$$ (5.2.22)

*c)* with respect to the origin $O$, i.e. ***the polar moment of inertia.***

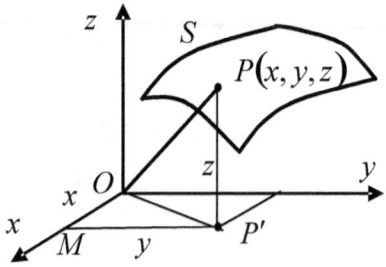

*Figure 5. 11. The distance of the current point on the surface to the origin*

In the figure 5.11, $\overline{OP}$ is the diagonal of the parallelepiped of sides $x, y, z$, having its left hind edge at the origin and its faces parallel to the coordinate planes. $\left|\overline{OP}\right| \equiv OP$ represents the distance from the current point $P$ to the origin. In the triangle $OPP'$ we have

$$OP^2 = OP'^2 + PP'^2 = OP'^2 + z^2, \qquad (5.2.23)$$

and in the triangle $MP'O$

$$OP'^2 = OM^2 + MP'^2 = x^2 + y^2. \qquad (5.2.24)$$

Consequently, $OP^2 = x^2 + y^2 + z^2$. The polar moment of inertia is given by

$$I_O = \iint_S \left(x^2 + y^2 + z^2\right)\gamma\left(x, y, z\right)\mathrm{d}\sigma. \qquad (5.2.25)$$

We notice that the following equalities are valid

$$I_O = I_x + I_{yz} = I_y + I_{zx} = I_z + I_{xy} = I_{xy} + I_{yz} + I_{zx}. \qquad (5.2.26)$$

Let us specify that the geometric quantities are defined by putting $\gamma = 1$ in the corresponding formulas.

*Examples*

1. Compute the area of the surface representing the intersection of the cone $z = \sqrt{x^2 + y^2}$ with the cylinder $x^2 + y^2 = 2x$, placed inside the cylinder (figure 5.12).

**Solution.**

The circle $D$ is the projection on $xOy$ of the above described surface $S$. Hence, the area of $S$ is computed by means of the formula

$$|S| = \iint_S \mathrm{d}\sigma = \iint_D \sqrt{1 + \left(\frac{\partial z}{\partial x}\right)^2 + \left(\frac{\partial z}{\partial y}\right)^2}\,\mathrm{d}x\,\mathrm{d}y. \qquad (5.2.27)$$

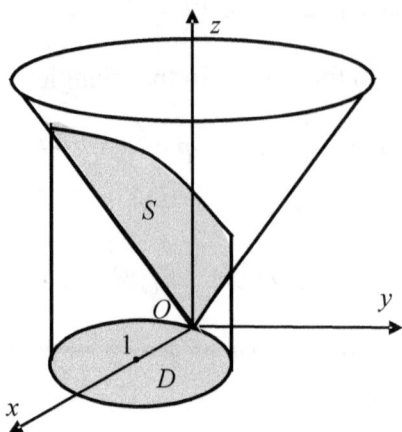

*Figure 5. 12. The surface S, representing intersection between the cone of equation*

$$z = \sqrt{x^2 + y^2} \text{ and the cylinder } x^2 + y^2 = 2x$$

Let us calculate the partial derivatives under the radical. We have

$$\frac{\partial z}{\partial x} = \frac{\partial}{\partial x}\left(\sqrt{x^2 + y^2}\right) = \frac{x}{\sqrt{x^2 + y^2}},$$

$$\frac{\partial z}{\partial y} = \frac{\partial}{\partial y}\left(\sqrt{x^2 + y^2}\right) = \frac{y}{\sqrt{x^2 + y^2}},$$

(5.2.28)

hence,

$$\left(\frac{\partial z}{\partial x}\right)^2 + \left(\frac{\partial z}{\partial y}\right)^2 = 1.$$

(5.2.29)

We get

$$|S| = \iint_D \sqrt{2}\, dx\, dy = \sqrt{2} \iint_D dx\, dy.$$

(5.2.30)

There are two possibilities for the calculation of the double integral:

a) The easiest one, considering that $|D| = \iint\limits_D dx\, dy$. The area of

the circle is $\pi \cdot R^2 = \pi \cdot 1^2 = \pi$.

b) The more complicated one, but emphasizing a good training for such calculations: we compute the double integral by polar coordinates

$$T : \begin{cases} x = \rho\cos\theta \\ y = \rho\sin\theta \end{cases}, \quad \frac{D(x,y)}{D(\rho,\theta)} = \rho, \text{ therefore}$$

$$\iint\limits_D dx\, dy = \iint\limits_9 \rho\, d\rho\, d\theta .$$

(5.2.31)

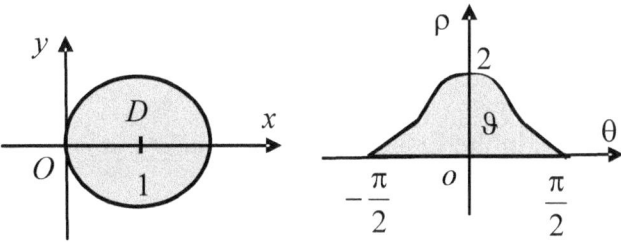

*Figure 5. 13. The transformation of the domain D in polar coordinates*

The domain $9$ from the plane $\rho o\theta$ is simple $o\theta$ and it is bounded by the curve obtained by writing the equation of the cylinder in polar coordinates. We have

$$\rho^2 = 2\rho\cos\theta ,$$

(5.2.32)

i.e.

$$\rho = 2\cos\theta .$$

(5.2.33)

Consequently,

$$|D| = \int\limits_{-\frac{\pi}{2}}^{\frac{\pi}{2}} d\theta \int\limits_{0}^{2\cos\theta} \rho\,d\rho = \int\limits_{-\frac{\pi}{2}}^{\frac{\pi}{2}} \frac{\rho^2}{2}\bigg|_{\rho=0}^{\rho=2\cos\theta} d\theta =$$

$$= 2\int\limits_{-\frac{\pi}{2}}^{\frac{\pi}{2}} \cos^2\theta\,d\theta = \int\limits_{-\frac{\pi}{2}}^{\frac{\pi}{2}} (1+\cos 2\theta)d\theta \Rightarrow |D| = \pi,$$

(5.2.34)

i.e.

$$\boxed{|S| = \pi\sqrt{2}}.$$

(5.2.35)

2. Compute the area of the portion of the sphere $x^2 + y^2 + z^2 = a^2$ enclosed in the cylinder $x^2 + y^2 = ay$ (the portion included in the first octant).

**Solution.** In the first octant, $z > 0$, hence $z = \sqrt{a^2 - x^2 - y^2}$.

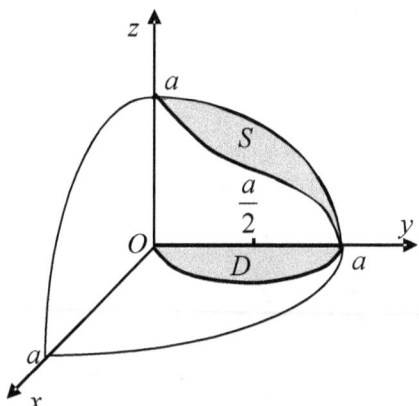

Figure 5. 14. The intersection between the sphere $x^2 + y^2 + z^2 = a^2$ and the cylinder $x^2 + y^2 = ay$

The projection of $S$ on the plane $xOy$ is the semicircle $x^2 + y^2 = ay$. The area of $S$ is thus

$$|S| = \iint_S d\sigma = \iint_D \sqrt{1 + p^2 + q^2}\, dx\, dy. \qquad (5.2.36)$$

But

$$p = \frac{\partial z}{\partial x} = -\frac{x}{\sqrt{a^2 - x^2 - y^2}},$$

$$q = \frac{\partial z}{\partial y} = -\frac{y}{\sqrt{a^2 - x^2 - y^2}}, \qquad (5.2.37)$$

whence

$$1 + p^2 + q^2 = 1 + \frac{x^2 + y^2}{a^2 - x^2 - y^2} = \frac{a^2}{a^2 - x^2 - y^2}. \qquad (5.2.38)$$

It remains to compute the double integral

$$|S| = \iint_D \frac{a}{\sqrt{a^2 - x^2 - y^2}}\, dx\, dy \qquad (5.2.39)$$

on the semicircle from figure 5.15 a).

Let us apply polar coordinates. We have

$$\begin{cases} x = \rho\cos\theta, \\ y = \rho\sin\theta, \end{cases} \quad \theta \in \left[0, \frac{\pi}{2}\right], \quad \rho \in [0, a], \qquad (5.2.40)$$

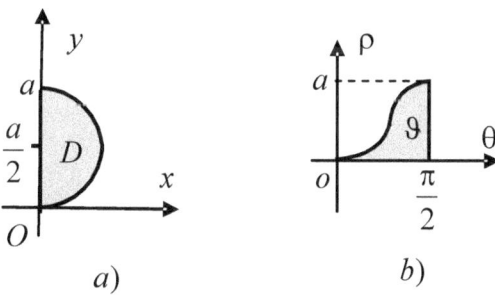

*a)*                   *b)*

*Figure 5. 15. The transformation of the semicircle D into the domain* $\vartheta$

hence

$$|S| = \iint_\vartheta \frac{a}{\sqrt{a^2 - \rho^2}} \cdot \rho\, d\rho\, d\theta, \qquad (5.2.41)$$

241

where $\left|\dfrac{D(x,y)}{D(\rho,\theta)}\right| = \rho$, and $\vartheta$ is the transformed domain from figure

5.15 b). The circle equation $x^2 + y^2 = ay$ reads $\rho^2 = a \cdot \rho \sin \theta$, which becomes, after simplification, $\rho = a \cdot \sin \theta$.

Step by step, it follows

$$|S| = \int_0^{\frac{\pi}{2}} d\theta \int_0^{a\sin\theta} \frac{a\rho}{\sqrt{a^2 - \rho^2}} d\rho = -\int_0^{\frac{\pi}{2}} a\sqrt{a^2 - \rho^2}\,\Big|_{\rho=0}^{\rho=a\sin\theta} d\theta =$$

$$= a\int_0^{\frac{\pi}{2}} (a - a\cos\theta)\,d\theta \Rightarrow \boxed{|S| = a^2\left(\frac{\pi}{2} - 1\right)}.$$

(5.2.42)

3. Find the geometric moment of inertia of the hemisphere $z = \sqrt{a^2 - x^2 - y^2}$ relative to the $Oz$ axis.

**Solution.**

$$I_z = \iint_S (x^2 + y^2)\,d\sigma = \iint_D (x^2 + y^2)\sqrt{1 + p^2 + q^2}\,dx\,dy,$$

$$\left.\begin{array}{l} p = \dfrac{\partial z}{\partial x} = -\dfrac{x}{\sqrt{a^2 - x^2 - y^2}} \\[3mm] q = \dfrac{\partial z}{\partial y} = -\dfrac{y}{\sqrt{a^2 - x^2 - y^2}} \end{array}\right| \Rightarrow 1 + p^2 + q^2 = \dfrac{a^2}{a^2 - x^2 - y^2}.$$

(5.2.43)

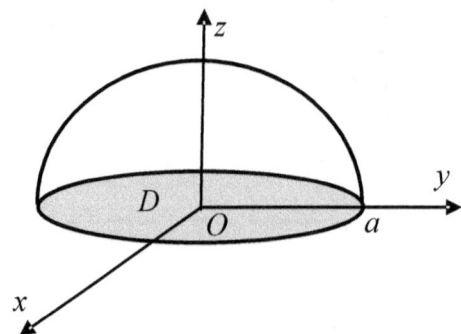

*Figure 5. 16. The projection of the hemisphere of radius a on the plane xOy*

242

Thus,

$$I_z = \iint\limits_D \left(x^2 + y^2\right) \cdot \frac{a}{\sqrt{a^2 - x^2 - y^2}} \, dx \, dy. \qquad (5.2.44)$$

Using polar coordinates $\begin{cases} x = \rho\cos\theta \\ y = \rho\sin\theta \end{cases}$, the domain $D$ from

figure 5.17 a) becomes the rectangle $\vartheta$, from 5. 17 b). It follows that

$$I_z = \iint\limits_{\vartheta} \frac{\rho^2 \cdot a}{\sqrt{a^2 - \rho^2}} \cdot \rho \, d\rho \, d\theta. \qquad (5.2.45)$$

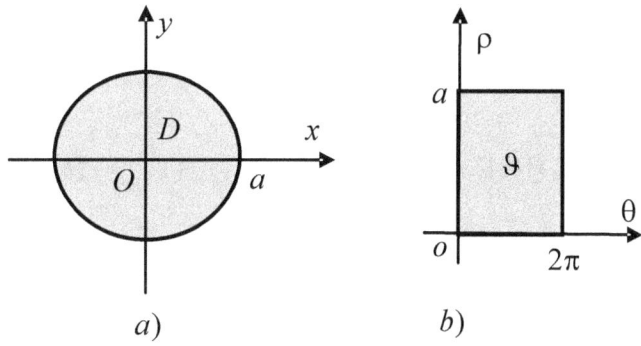

$a)$ $\qquad\qquad\qquad$ $b)$

*Figure 5. 17. The circle D is mapped into the rectangle $\vartheta$*

The domain $\vartheta$ is both simple $o\theta$ and $o\rho$. The integrand depends only on $\rho$, which leads to

$$I_z = \int\limits_0^{2\pi} d\theta \cdot \int\limits_0^a \frac{\rho^3 a}{\sqrt{a^2 - \rho^2}} \, d\rho = 2\pi a \int\limits_0^a \frac{\rho^3}{\sqrt{a^2 - \rho^2}} \, d\rho. \qquad (5.2.46)$$

In the simple integral from above, we make the change of variable $\rho = a\sin\theta$ and we obtain

$$I_z = 2\pi a \cdot \int\limits_0^{\frac{\pi}{2}} \frac{a^3 \sin^3 t}{a \cos t} \cdot \underbrace{a \cos t \, dt}_{d\rho} = 2\pi a^4 \cdot \int\limits_0^{\frac{\pi}{2}} \sin^3 t \, dt =$$

$$= 2\pi a^4 \cdot \int\limits_0^{\frac{\pi}{2}} \sin t \left(1 - \cos^2 t\right) dt = \hspace{2cm} (5.2.47)$$

$$= 2\pi a^4 \left(-\cos t + \frac{\cos^3 t}{3}\right)\Bigg|_{t=0}^{t=\frac{\pi}{2}}.$$

Finally,

$$\boxed{I_z = \frac{4\pi a^4}{3}}. \hspace{2cm} (5.2.48)$$

## 5.3. THE SURFACE INTEGRAL OF THE SECOND KIND

In order to define the second kind surface integral, it is necessary to bring forward firstly the notion of *orientation on surfaces*.

Let $S$ be an unclosed smooth surface with two faces and bounded by a closed simple curve $c$. Let us choose a face of this surface. We give the curve $c$ *the direct sense* (i.e. counterclockwise). Considering the same rule, we can establish, at the same time, the positive sense on each closed simple curve situated on the surface. All these define the *orientation* on that particular surface.

If we start from the other face of the surface, then the position of the observer changes, the normals change their sense and, according to the previously established rule, the sense on $c$, as well as the sense on the other closed curves situated on $S$, also changes.

Therefore, if we stick to this rule, ***the choice of a side of the surface defines its orientation*** and, reciprocally, ***the choice of the positive sense on a closed simple curve situated on that surface unambiguously defines its face.*** If a smooth closed surface $S$ is "the skin" of a body, this rule, obviously, works no more; we can talk about the exterior surface of the body, which has a ***positive orientation*** or about the interior one, which has a ***negative orientation***. If we consider the mathematical convention of using on each surface, as positive orientation, the one which corresponds to the exterior face of the surface and, as negative orientation, the opposite one, then we can build up an orientation defined on the entire space, as the choice of a positive sense on the curves of a plane characterizes the orientation of the plane. The space orientation to the right corresponds to the direct sense and that to the left – to the inverse sense. In order to avoid confusion, we shall assume that the space is oriented to the right. In order to characterize the chosen face of a surface, we shall use the sign + before the radical in the formulas of the direction cosines of the outward normal to the surface.

The second kind surface integral is defined analogously to the curvilinear integral of the same kind. There, we started from an oriented curve on which a partition was set up and we projected it on the axes of coordinates, keeping the sense on the curve. The obtained projection also kept the same sense.

Similarly, we consider a smooth – or picewisely smooth – surface with two faces and we choose a certain orientation on it.

Firstly, we assume that the surface is explicitly given, i.e.

$$z = z(x, y) \qquad (5.3.1)$$

and that the point $(x, y)$ varies in the domain $D$ – the projection of the surface on the plane $xOy$. An arbitrary partition of the surface $\Delta = \{S_1, S_2, \ldots, S_n\}$ (figure 5.7) will also be projected conserving the orientation; the projection of each element $S_j$ of the partition $\Delta$ to the plane $xOy$ is a domain $D_j$ situated in the plane, and, obviously, $\delta = \{D_1, D_2, \ldots, D_n\}$ is a partition for $D$. The **norm** of the partition $\Delta$ is $\nu(\Delta) = \max \{\operatorname{diam} S_j, \ j = \overline{1, n}\}$. Now, consider a function $f = f(x, y, z)$ defined on this surface, i.e. $f : S \to \Re$. Let us choose on each $S_j$ a point $M_j(x_j, y_j, z_j) \in S_j$, $j = \overline{1, n}$. Denoting the area of $D_j$ by $\Delta D_j$, we set up the Riemann sum

$$\sigma_{\Delta}(f, M_j) = \sum_{j=1}^{n} f(x_j, y_j, z_j) \Delta D_j . \qquad (5.3.2)$$

The finite limit of this sum (if it exists!), when the norm tends to zero for any choice of the intermediate points $M_j$, is called **the surface integral of the second kind** of the function $f$ and it is denoted by

$$I = \iint_S f(x, y, z) \, dx \, dy . \qquad (5.3.3)$$

*Remark.* $dx \, dy$ suggests the area of the projection of a surface element on the plane $xOy$.

If we project the surface elements on the planes $yOz$ or $zOx$, instead of projecting them on the plane $xOy$, we obtain other two second kind surface integrals

$$\iint_S f(x,y,z)\mathrm{d}y\,\mathrm{d}z, \quad \iint_S f(x,y,z)\mathrm{d}z\,\mathrm{d}x. \qquad (5.3.4)$$

In applications, we often find integral combinations of all of these forms

$$I = \iint_S P(x,y,z)\mathrm{d}y\,\mathrm{d}z + Q(x,y,z)\mathrm{d}z\,\mathrm{d}x + R(x,y,z)\mathrm{d}x\,\mathrm{d}y, \qquad (5.3.5)$$

where $P, Q, R : S \to \Re$ are, usually, of class $C^0(S)$.

It should be mentioned that, in all cases, we assume that *the surface S has two faces and the integral is applied only to one of them.*

The integral $I$ can easily be reduced to a first kind surface integral. We can emphasize this fact considering the figure 5.18.

Let us set up the tangent plane $(p)$ at $P \in \Delta S$ to $S$ by plotting the normal to the surface at the current point $P(x,y,z)$. Let $(\alpha, \beta, \gamma)$ be the direction cosines of the normal to $S$. The angle between the unit vector **n** of the outward normal to the surface and $Oz$ is equal to the dihedral angle between the plane $xOy$ and the plane $(p)$, as they are angles with perpendicular sides (the angles denoted by $\theta$ in the figure 5.18).

The projection of the surface element of area $\mathrm{d}\sigma$ on the plane $xOy$ is the rectangle of sides $\mathrm{d}x, \mathrm{d}y$. Here we apply the cosine formula, which states that the area of the projected surface is equal to the area of the surface which is projected multiplied by the cosine of the dihedral angle between the projection planes.

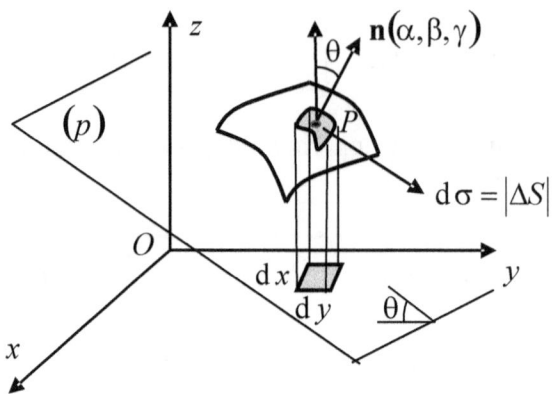

*Figure 5. 18. The surface element of area* $\mathrm{d}\sigma$

In this case, the element of surface of area $\mathrm{d}\sigma$ can be regarded as a plane surface; we can make this assumption for $\mathrm{d}\sigma$ small enough, i.e. the surface element is included in the plane $(p)$ tangent to $S$. We obtain

$$
\begin{aligned}
\mathrm{d}x\,\mathrm{d}y &= \gamma\,\mathrm{d}\sigma, \\
\mathrm{d}y\,\mathrm{d}z &= \alpha\,\mathrm{d}\sigma, \\
\mathrm{d}z\,\mathrm{d}x &= \beta\,\mathrm{d}\sigma,
\end{aligned}
\qquad (5.3.6)
$$

and from here it results

$$
I = \iint\limits_{S} (P\alpha + Q\beta + R\gamma)\,\mathrm{d}\sigma, \qquad (5.3.7)
$$

which is a first kind surface integral, of integrand $P\alpha + Q\beta + R\gamma$.

This integrand can also be written otherwise. If $\mathbf{V}(P,Q,R): S \to \mathfrak{R}^{3}$, then $\mathbf{V} \cdot \mathbf{n} = P\alpha + Q\beta + R\gamma$. Consequently,

$$I = \iint_S \mathbf{V} \cdot \mathbf{n} \, d\sigma . \qquad (5.3.8)$$

If we write it in this form, the surface integral of the second kind emphasizes the following

### GEOMETRIC INTERPRETATION

Let $d\sigma$ be the area of the surface element and $\mathbf{n}$ be the normal to $S$ in $P$ (figure 5.19).

Then $\mathbf{V} \cdot \mathbf{n}$ is the projection of $\mathbf{V}$ on the direction of the normal to the surface, hence $\mathbf{V} \cdot \mathbf{n} = PP'$. The product $PP' \cdot d\sigma$ is the volume of the right cylinder having the area of basis equal to $d\sigma$ and the height $PP'$. If we denote this volume by $v$, we have

$$v = \mathbf{V} \cdot \mathbf{n} \, d\sigma . \qquad (5.3.9)$$

This volume is also called the ***elementary flow*** of the vector $\mathbf{V}$ through the element $d\sigma$.

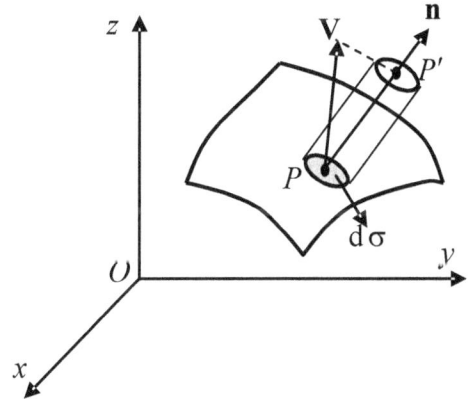

*Figure 5. 19. The elementary flow*

The denomination comes from hydrodynamics, i.e. if $\mathbf{V}$ is the velocity of a fluid, then $\mathbf{V} \cdot \mathbf{n} \, d\sigma$ measures the quantity of fluid which passes through $d\sigma$ in the time unit (in one sense or another, considering the sign of $\mathbf{V} \cdot \mathbf{n}$).

It follows that

$$I = \iint_S \mathbf{V} \cdot \mathbf{n} \, d\sigma \qquad (5.3.10)$$

is **the total flow** which passes through the surface $S$.

This is **the main physical interpretation** of the surface integral of the second kind.

*Example.* Find the flow of the vector $\mathbf{r} = x\mathbf{i} + y\mathbf{j} + z\mathbf{k}$ through the exterior surface of the right circular cylinder with $Oz$ as axis of symmetry, having the basis of radius $R$ situated in plane $xOy$ and the height $h$.

**Solution.** According to the above mentioned physical interpretation, the flow is computed using the formula

$$\Phi = \iint_S \mathbf{r} \cdot \mathbf{n} \, d\sigma, \qquad S = S_1 \cup S_2 \cup S_3. \qquad (5.3.11)$$

If we apply the property of additivity of the integral with respect to the domain of integration, we infer that

$$\Phi = \underbrace{\iint_{S_1} \mathbf{r} \cdot \mathbf{n} \, d\sigma}_{\Phi_1} + \underbrace{\iint_{S_2} \mathbf{r} \cdot \mathbf{n} \, d\sigma}_{\Phi_2} + \underbrace{\iint_{S_3} \mathbf{r} \cdot \mathbf{n} \, d\sigma}_{\Phi_3}. \qquad (5.3.12)$$

We compute, by turns, $\Phi_1, \Phi_2, \Phi_3$.

On $S_1$, $\mathbf{r}$ is in the plane $xOy$. The normal to $S_1$ is perpendicular to $xOy$, therefore $\mathbf{r} \cdot \mathbf{n} = 0$, whence we obtain

$$\Phi_1 = 0. \qquad (5.3.13)$$

On $S_3$, **r** joins the origin with an arbitrary point from the surface of the cylinder.

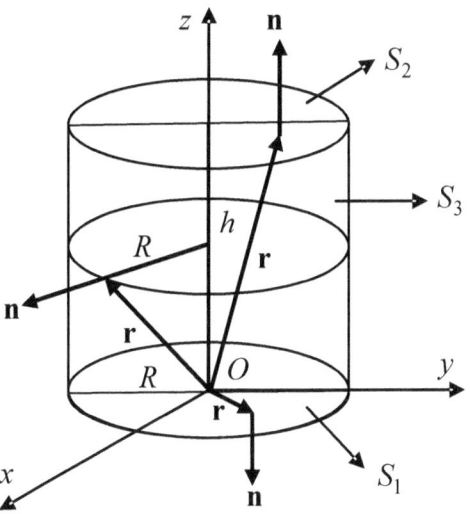

*Figure 5. 20. The right circular cylinder*

The normal to $S_3$ is thus parallel to $xOy$ and $\mathbf{r} \cdot \mathbf{n} = R$, for any $P \in S_3$. We get

$$\Phi_3 = \iint_{S_3} R \, d\sigma = R \cdot \iint_{S_3} d\sigma = R \left| S_3 \right|, \tag{5.3.14}$$

where $\left| S_3 \right|$ is the area of $S_3$. But the lateral area of the circular cylinder is $\left| S_3 \right| = 2\pi R h$, hence

$$\Phi_3 = 2\pi R^2 h. \tag{5.3.15}$$

On $S_2$, **n** is perpendicular to $Oz$, therefore its projection on $Oz$ has the length $h$, for any $P \in S_2$. It follows that

$$\Phi_2 = \iint_{S_2} h \, d\sigma = h \cdot \iint_{S_2} d\sigma = h \left| S_2 \right|. \tag{5.3.16}$$

But $\left|S_2\right| = \pi R^2$, which yields

$$\Phi_2 = \pi R^2 h. \tag{5.3.17}$$

Finally, by adding the values of the three fluxes, we obtain

$$\Phi = \Phi_1 + \Phi_2 + \Phi_3 = 0 + \pi R^2 h + 2\pi R^2 h, \tag{5.3.18}$$

and in conclusion

$$\boxed{\Phi = 3\pi R^2 h}. \tag{5.3.19}$$

## 5.4. INTEGRAL FORMULAS

We shall describe two of these types of formulas: the Stokes formula and the flux-divergence formula (or the Gauss-Ostrogradsky formula), which are the most significant and useful in applications.

### 5.4.1. STOKES' FORMULA

We recall from chapter 3 **Green's formula**, which turns a curvilinear integral taken along a closed curve in plane into a double integral on the domain bounded by this curve:

$$\oint_\Gamma P(x,y)\,dx + Q(x,y)\,dy = \iint_D \left( \frac{\partial Q}{\partial x} - \frac{\partial P}{\partial y} \right) dx\,dy. \tag{5.4.1}$$

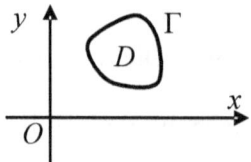

*Figure 5. 21. The domain D of boundary* $\Gamma$

**PROBLEM.** Is it possible to do the same thing on a deformed domain, which becomes a portion of surface, bounded by a closed curve? (figure 5.22).

The answer is affirmative. We obtain **Stokes' formula.**

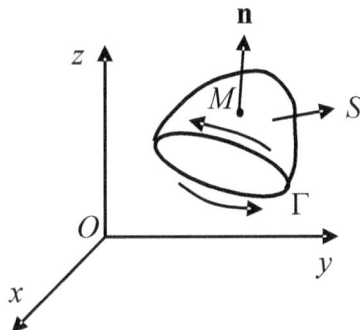

*Figure 5. 22. The domain S bounded by the curve Γ*

Let $P, Q, R \in C^1(S)$ and consider the curvilinear integral

$$\oint_{\Gamma} P\,dx + Q\,dy + R\,dz, \tag{5.4.2}$$

which can also be written in vector form as $\oint_{\Gamma} \mathbf{V} \cdot d\mathbf{r}$, where

$\mathbf{V}(P, Q, R) \in \left(C^1(S)\right)^3$, and $d\mathbf{r} = \mathbf{i}\,dx + \mathbf{j}\,dy + \mathbf{k}\,dz$.

Then, we can prove that

$$\oint_{\Gamma} \mathbf{V} \cdot d\mathbf{r} = \iint_{S} \mathrm{rot}\,\mathbf{V} \cdot \mathbf{n}\,d\sigma. \tag{5.4.3}$$

This is **Stokes' formula.**

### PHYSICAL INTERPRETATION

**The circulation of V along the closed curve Γ is equal to the flow of the curl of V through any surface leaning on Γ.**

To develop Stokes' formula, we compute $\oint_\Gamma \mathbf{V} \cdot d\mathbf{r} = \iint_S \operatorname{rot} \mathbf{V} \cdot \mathbf{n} \, d\sigma$ using the well-known formal determinant:

$$\operatorname{rot} \mathbf{V} = \begin{vmatrix} \mathbf{i} & \mathbf{j} & \mathbf{k} \\ \dfrac{\partial}{\partial x} & \dfrac{\partial}{\partial y} & \dfrac{\partial}{\partial z} \\ P & Q & R \end{vmatrix} = \tag{5.4.4}$$

$$= \mathbf{i}\left(\frac{\partial R}{\partial y} - \frac{\partial Q}{\partial z}\right) + \mathbf{j}\left(\frac{\partial P}{\partial z} - \frac{\partial R}{\partial x}\right) + \mathbf{k}\left(\frac{\partial Q}{\partial x} - \frac{\partial P}{\partial y}\right).$$

We obtain

$$\oint_\Gamma P \, dx + Q \, dy + R \, dz = \iint_S \left[ \left(\frac{\partial R}{\partial y} - \frac{\partial Q}{\partial z}\right)\alpha + \left(\frac{\partial P}{\partial z} - \frac{\partial R}{\partial x}\right)\beta + \right.$$

$$\left. + \left(\frac{\partial Q}{\partial x} - \frac{\partial P}{\partial y}\right)\gamma \right] d\sigma, \tag{5.4.5}$$

where $\alpha$, $\beta$, $\delta$ are the direction cosines of the outward normal to $S$.

But, as we have previously shown, $\alpha \, d\sigma = dy \, dz$, $\beta \, d\sigma = dx \, dz$, $\gamma \, d\sigma = dx \, dy$, hence Stokes' formula can be also written in the form

$$\oint_\Gamma P \, dx + Q \, dy + R \, dz =$$

$$= \iint_S \left(\frac{\partial R}{\partial y} - \frac{\partial Q}{\partial z}\right) dy \, dz + \left(\frac{\partial P}{\partial z} - \frac{\partial R}{\partial x}\right) dx \, dz + \left(\frac{\partial Q}{\partial x} - \frac{\partial P}{\partial y}\right) dx \, dy. \tag{5.4.6}$$

It is a ***generalization of Green's formula***.

Indeed, if $S$ is a flat surface included in $xOy$, for example, then we have the same figure as for Green's formula: $S$ becomes $D$, bounded by the closed curve $\Gamma$.

The normal to the plane $xOy$ (i.e. to $S$) has the direction cosines $(0,0,1)$. Therefore, if $\mathbf{V} = \mathbf{V}\big(P(x,y),Q(x,y),0\big)$ is a plane vector field, we have

$$\mathrm{rot}\,\mathbf{V}\cdot\mathbf{n} = 0\cdot\left(\frac{\partial R}{\partial y} - \frac{\partial Q}{\partial z}\right) + 0\cdot\left(\frac{\partial P}{\partial z} - \frac{\partial R}{\partial x}\right) + 1\cdot\left(\frac{\partial Q}{\partial x} - \frac{\partial P}{\partial y}\right), \quad (5.4.7)$$

which yields

$$\mathrm{rot}\,\mathbf{V}\cdot\mathbf{n} = \frac{\partial Q}{\partial x} - \frac{\partial P}{\partial y}. \tag{5.4.8}$$

Thus, the curvilinear integral becomes $\oint_{\Gamma} P\,dx + Q\,dy$ and we obtain Green's formula.

Hence, **Green's formula is a particular case of Stokes' formula**.

**APPLICATION.** Let $\mathbf{V}(P,Q,R)$ be irrotational, with $P,Q,R$ of class $C^1$ on the three-dimensional domain $\Omega$. This means that $\mathrm{rot}\,\mathbf{V} = \mathbf{0}$ in $\Omega$.

Now, let $\Gamma \subset \Omega$ be a closed curve and $S$ any surface also included in $\Omega$ and leaning on $\Gamma$. Applying Stokes' formula, we get:

$$\oint_{\Gamma} \mathbf{V}\cdot d\mathbf{r} = \iint_{S} \underbrace{\mathrm{rot}\,\mathbf{V}\cdot\mathbf{n}}_{\equiv 0}\, d\sigma = 0 \quad \Rightarrow \quad \boxed{\oint_{\Gamma} \mathbf{V}\cdot d\mathbf{r} = 0}. \tag{5.4.9}$$

In conclusion, if $\mathbf{V}$ is an irrotational field, **the circulation along any closed curve $\Gamma \subset \Omega$ is null**.

*Example:* Find the circulation of the vector $\mathbf{V}(y+z,\ z+x,\ x+y)$ along the curve $C$, which is the intersection between the sphere $x^2 + y^2 + z^2 = a^2$ and the plane $x + y + z = 0$.

**Solution.** According to the ennounce, the components of **V** are

$$P = y + z, \quad Q = z + x, \quad R = x + y. \tag{5.4.10}$$

We compute the circulation using the formula $c = \oint_C \mathbf{V} \cdot d\mathbf{r}$ , i.e.

$$c = \oint_C (y + z)\,dx + (x + z)\,dy + (x + y)\,dz. \tag{5.4.11}$$

We can compute it directly, but it is easier to apply Stokes' formula:

$$c = \oint_C \mathbf{V} \cdot d\mathbf{r} = \iint_S \operatorname{rot} \mathbf{V} \cdot \mathbf{n}\; d\sigma. \tag{5.4.12}$$

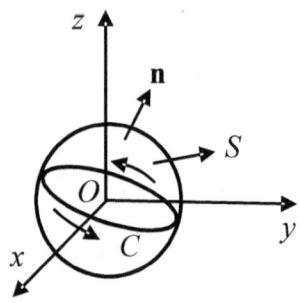

*Figure 5. 23. The intersection between the sphere of equation* $x^2 + y^2 + z^2 = a^2$
*and the plane* $x + y + z = 0$

We have

$$\operatorname{rot} \mathbf{V} = \begin{vmatrix} \mathbf{i} & \mathbf{j} & \mathbf{k} \\ \dfrac{\partial}{\partial x} & \dfrac{\partial}{\partial y} & \dfrac{\partial}{\partial z} \\ y+z & x+z & x+y \end{vmatrix} = \mathbf{i}(1-1) - \mathbf{j}(1-1) + \mathbf{k}(1-1); \tag{5.4.13}$$

we obtain $\operatorname{rot} \mathbf{V} = \mathbf{0}$, hence $\boxed{c = 0}$ , because the field is irrotational.

## 5.4.2. THE FLUX-DIVERGENCE FORMULA

Green's and Stokes' formulas are able to transform certain double or surface integrals into curvilinear integrals.

The flux-divergence formula (or Gauss-Ostrogradsky's formula) is similar and it allows the transformation of some triple integrals into surface integrals. In fact, all of these formulas – Green, Stokes, flux-divergence – are extensions of the fundamental formula for defined integrals:

$$\int_a^b f'(x)\,dx = f(b) - f(a), \qquad (5.4.14)$$

which reduces the calculus of an integral to a calculus on the boundary of the domain of integration – in formula (5.4.14), this boundary is represented by the endpoints of the interval $[a, b]$.

Let the domain $\Omega \subset \Re^3$ be simple with respect to the axis of coordinates. Let $P, Q, R \in C^1(\Omega)$.

Consider the triple integral $\iiint_\Omega \left( \dfrac{\partial P}{\partial x} + \dfrac{\partial Q}{\partial y} + \dfrac{\partial R}{\partial z} \right) dx\, dy\, dz$.

Let us compute the integrals from each of the terms from above, by turns.

As $\Omega$ is simple $Oz$, we have

$$\iiint_\Omega \frac{\partial R}{\partial z}\,dx\,dy\,dz = \iint_D dx\,dy \int_{z_1}^{z_2} \frac{\partial R}{\partial z}\,dz. \qquad (5.4.15)$$

Let us assume that $S = S_1^- \cup S_2$, where $S = fr\,\Omega$ ($S_1^-$ is for orientation), and $S_1, S_2$ are expressed by

257

$$(S_1): z = \varphi_1(x, y), \quad (S_2): z = \varphi_2(x, y). \tag{5.4.16}$$

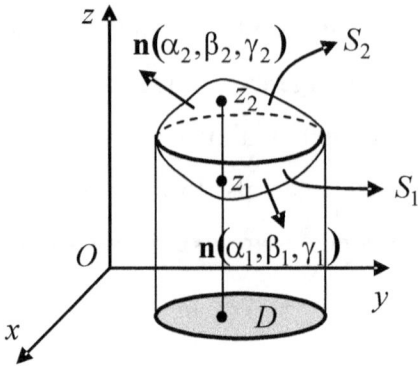

*Figure 5. 24. Domain simple with respect to the axes of coordinates*

Hence,

$$\iiint\limits_{\Omega} \frac{\partial R}{\partial z} \, dx \, dy \, dz = \iint\limits_{D} R(x, y, z) \Big|_{z=\varphi_1(x,y)}^{z=\varphi_2(x,y)} \, dx \, dy =$$

$$= \iint\limits_{D} R(x, y, \varphi_2(x, y)) \, dx \, dy -$$

$$- \iint\limits_{D} R(x, y, \varphi_1(x, y)) \, dx \, dy = \tag{5.4.17}$$

$$= \iint\limits_{S_2} R(x, y, z_2) \gamma_2 \, dS_2 - \iint\limits_{S_1} R(x, y, z_1) \gamma_1 \, dS_1 =$$

$$= \iint\limits_{S} R(x, y, z) \gamma \, dS.$$

Analogously, if we project $\Omega$ on the other coordinate planes, we get similar formulas for the other two integrals. Therefore,

$$\iiint\limits_{\Omega} \frac{\partial P}{\partial x} \, dx \, dy \, dz = \iint\limits_{S} P \alpha \, dS,$$

$$\iiint\limits_{\Omega} \frac{\partial Q}{\partial y} \, dx \, dy \, dz = \iint\limits_{S} Q \beta \, dS, \tag{5.4.18}$$

$$\iiint\limits_{\Omega} \frac{\partial R}{\partial z} \, dx \, dy \, dz = \iint\limits_{S} R \gamma \, dS.$$

If we add the three formulas member by member, then we obtain **the flux-divergence formula (Gauss-Ostrogradsky's formula):**

$$\iiint_{\Omega}\left(\frac{\partial P}{\partial x} + \frac{\partial Q}{\partial y} + \frac{\partial R}{\partial z}\right) dx\, dy\, dz = \iint_{S}(P\alpha + Q\beta + R\gamma)dS. \qquad (5.4.19)$$

### PHYSICAL INTERPRETATION

If $V(P, Q, R) \in \left[C^{1}(\Omega)\right]^{3}$ is a field of velocities, then

$$\operatorname{div} \mathbf{V} = \nabla \cdot \mathbf{V} = \frac{\partial P}{\partial x} + \frac{\partial Q}{\partial y} + \frac{\partial R}{\partial z},$$
$$(5.4.20)$$
$$\mathbf{V} \cdot \mathbf{n} = P\alpha + Q\beta + R\gamma,$$

and formula (5.4.19) can be also written as

$$\iiint_{\Omega} \operatorname{div} \mathbf{V}\, dx\, dy\, dz = \iint_{S} \mathbf{V} \cdot \mathbf{n}\, dS. \qquad (5.4.21)$$

In the triple integral, the integrand is $\operatorname{div} \mathbf{V}$, and the surface integral on the right side is actually the total flow of $\mathbf{V}$ through the closed surface $S$.

**PARTICULAR CASE:** If $\mathbf{V}$ is a solenoidal field, then $\operatorname{div} \mathbf{V} = 0$ in $\Omega$ and it follows that the triple integral cancels identically, on any subdomain $\Omega_{1} \subset \Omega$, of boundary $S_{1}$.

It follows that

$$\iint_{S_{1}} \mathbf{V} \cdot \mathbf{n}\, dS_{1} = 0, \qquad (5.4.22)$$

for any closed surface $S_{1} \subset \Omega$.

We conclude that

*In a solenoidal vector field, the total flow through any closed surface is null.*

*Example*: Compute the flow of the vector $\mathbf{V}(P,Q,R)$, $P = x^2 yz$, $Q = xy^2 z$, $R = xyz^2$, through the surface of the spherical triangle $x^2 + y^2 + z^2 = a^2$, $x = 0$, $y = 0$, $z = 0$.

**Solution.** The flow of $\mathbf{V}$ through $S$ is

$$\Phi = \iint\limits_S \mathbf{V} \cdot \mathbf{n}\, dS = \iint\limits_S \left(x^2 yz\alpha + xy^2 z\beta + xyz^2\gamma\right) dS.$$

According to the flux-divergence formula,

$$\Phi = \iiint\limits_\Omega \operatorname{div}\mathbf{V}\, dx\, dy\, dz. \tag{5.4.23}$$

We compute $\operatorname{div}\mathbf{V}$:

$$\operatorname{div}\mathbf{V} = \frac{\partial}{\partial x}\left(x^2 yz\right) + \frac{\partial}{\partial y}\left(xy^2 z\right) + \frac{\partial}{\partial z}\left(xyz^2\right) = 6xyz. \tag{5.4.24}$$

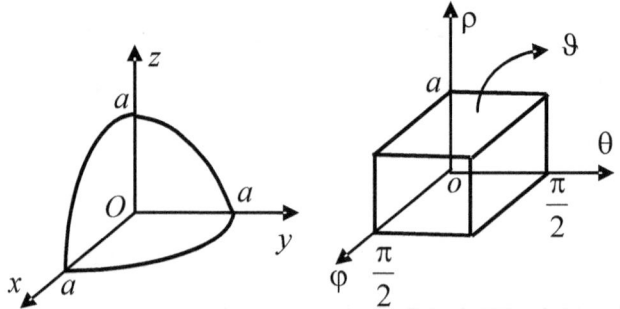

*Figure 5. 25. The transformation of the spherical triangle*

Therefore,

$$\Phi = 6\iiint\limits_\Omega xyz\, dx\, dy\, dz. \tag{5.4.25}$$

We apply now spherical coordinates:

260

$$\mathbf{T}: \begin{cases} x = \rho \cos\varphi \sin\theta \\ y = \rho \sin\varphi \sin\theta, \\ z = \rho \cos\theta \end{cases} \quad \rho \in [0, a], \ \theta \in \left[0, \frac{\pi}{2}\right],$$

(5.4.26)

$$\varphi \in \left[0, \frac{\pi}{2}\right].$$

We already know that

$$\left| \frac{D(x, y, z)}{D(\alpha, \beta, \gamma)} \right| = \rho^2 \sin\theta.$$

(5.4.27)

Let us compute the integrand with respect to spherical coordinates. We have

$$x\,y\,z = \rho \cos\varphi \sin\theta \cdot \rho \sin\varphi \sin\theta \cdot \rho \cos\theta =$$
$$= \rho^3 \cdot \cos\varphi \sin\varphi \cdot \sin^2\theta \cos\theta.$$

(5.4.28)

Replacing this in (5.4.25), we obtain

$$\Phi = \iiint_\vartheta \rho^3 \cdot \cos\varphi \sin\varphi \cdot \sin^2\theta \cos\theta \cdot \rho^2 \sin\theta \, d\rho \, d\theta \, d\varphi.$$

(5.4.29)

In the integral (5.4.29), $\vartheta$ is the parallelepiped from figure 5.25, and the variables separate within the integrand, hence the integral is computed as a product of three simple integrals:

$$\Phi = 6 \int_0^a \rho^5 \, d\rho \cdot \int_0^{\frac{\pi}{2}} \sin^3\theta \cos\theta \, d\theta \cdot \int_0^{\frac{\pi}{2}} \cos\varphi \sin\varphi \, d\varphi =$$

(5.4.30)

$$= 6 \frac{\rho^6}{6} \Big|_0^a \cdot \frac{\sin^4\theta}{4} \Big|_0^{\frac{\pi}{2}} \cdot \frac{1}{2} \left(\sin^2\varphi\right) \Big|_0^{\frac{\pi}{2}},$$

and, finally, it follows that

$$\boxed{\Phi = \frac{a^6}{8}}.$$

(5.4.31)

# EXERCISES AND PROBLEMS

## *The first kind surface integral*

1.  Compute the area of the upper hemisphere $x^2 + y^2 + z^2 = R^2, z \geq 0$.

$$\text{A: } |S| = 2\pi R^2$$

2. Compute the area of the section of surface of the paraboloid of equation $x^2 + y^2 = 2z$, bounded by the plane $z = 2$.

$$\text{A: } |S| = \frac{2\pi}{3}\left(5\sqrt{5} - 1\right)$$

3. Compute the area of the section of surface of the cone of equation $4\left(x^2 + y^2\right) - z^2 = 0$, bounded by the planes $z = 0$ and $z = 2$.

$$\text{A: } |S| = \pi\sqrt{5}$$

4. Find the area of the surface $y + z = a$, $x \geq 0$, $y \geq 0$, $z \geq 0$.

$$\text{A: } |S| = \frac{a^2\sqrt{3}}{2}$$

5. Find the moment of inertia relative to the $Oz$ axis of the cylinder $x^2 + y^2 = a$, with $0 \leq z \leq h$.

$$\text{A: } I_z = 2a^3 h\pi$$

6. Find the elevation of the geometric center of mass of the cone surface $z = \sqrt{x^2 + y^2}$, with $0 \leq z \leq 1$.

$$\text{A: } \bar{\bar{z}} = \frac{2}{3}$$

7. Find the moment of inertia relative to the $Oz$ axis of the lateral surface of the cone $z = \sqrt{x^2 + y^2}$, with $0 \le z \le 1$.

$$A:\ I_z = \frac{\sqrt{2}\pi}{2}$$

8. Compute $I = \iint\limits_S (x + y + z)\,d\sigma$, where $S$ is the surface of the hemisphere $x^2 + y^2 + z^2 = R^2, z \ge 0$.

$$A:\ I = \pi R^3$$

### The surface integral of the second kind

1. Compute $I = \iint\limits_S x\,dy\,dz + y\,dz\,dx + z\,dx\,dy$, where $S$ is the exterior face of the sphere $x^2 + y^2 + z^2 = R^2$.

$$A:\ I = 4\pi R^3$$

2. Compute

$$I = \iint\limits_S (y - z)\,dy\,dz + (z - x)\,dz\,dx + (x - y)\,dx\,dy,$$

where $S$ is the exterior face of the cone $x^2 + y^2 = z^2$, with $0 \le z \le h$.

$$A:\ I = 0$$

3. Compute the flow of the vector $\mathbf{r} = x\mathbf{i} + y\mathbf{j} + z\mathbf{k}$ through the exterior surface of the sphere of equation $x^2 + y^2 + z^2 = a^2$.

$$A:\ \Phi = 4\pi a^3$$

4. Compute the flow of the vector $\mathbf{r} = \dfrac{1}{x}\mathbf{i} + \dfrac{1}{y}\mathbf{j} + \dfrac{1}{z}\mathbf{k}$ through the exterior surface of the sphere of equation $x^2 + y^2 + z^2 = a^2$.

$$A:\ \Phi = 12\pi a^2$$

5. Compute the flow of the vector $\mathbf{V} = yz\mathbf{i} + zx\mathbf{j} + xy\mathbf{k}$ through the exterior surface of the cylinder $x^2 + y^2 = a^2$, with $0 \le z \le h$.

$$\text{A: } \Phi = 0$$

6. Compute the flow of the vector $\mathbf{V} = x^2\mathbf{i} + y^2\mathbf{j} + z^2\mathbf{k}$ through the exterior surface of the tetrahedron $OABC$, where $O(0,0,0)$, $A(1,0,0)$, $B(0,1,0)$, $C(0,0,1)$.

$$\text{A: } \Phi = \frac{1}{4}$$

7. Compute the flow of the vector $\mathbf{V} = x^3\mathbf{i} + y^3\mathbf{j} + z^3\mathbf{k}$ through the lateral surface of the cone $x^2 + y^2 = \dfrac{R^2}{h^2}z^2$, $0 \le z \le h$.

$$\text{A: } \Phi = \frac{\pi R^2 h}{10}\left(3R^2 - 4h^2\right)$$

## STOKES' FORMULA

1. Compute the circulation of the vector $\mathbf{V}(-y, x, c)$, $c \in \Re$, along the circle $C_1$ defined by the equations $x^2 + y^2 = 1$, $z = 0$, and along the circle $C_2$ defined by the equations $(x-2)^2 + y^2 = 1$, $z = 0$.

$$\text{A: } c_1 = c_2 = 2\pi$$

2. Compute the circulation of the vector $\mathbf{V}(0, x, y)$ along the curve $C$, which represents the intersection of the sphere $x^2 + y^2 + z^2 = a^2$ with the plane $x + y + z = 0$, crossed over in a positive sense.

$$A: c = \frac{\pi a^2 2\sqrt{3}}{3}$$

3. Compute the circulation of the vector $\mathbf{V}(y, z, x)$ along the curve $C$, which represents the intersection of the sphere $x^2 + y^2 + z^2 = a^2$ with the plane $x + z = a$.

$$A: c = -\frac{\pi a^2 \sqrt{2}}{2}$$

### THE FLUX-DIVERGENCE FORMULA
### (GAUSS-OSTROGRADSKY'S)

1. Compute the flow of the vector $\mathbf{r} = x\mathbf{i} + y\mathbf{j} + z\mathbf{k}$ through the exterior surface of the right circular cylinder $x^2 + y^2 = a^2$ bounded by the planes $z = -h, z = h$.

$$A: \Phi = 4\pi a^2 h$$

2. Compute the flow of the vector $\mathbf{V} = x^3\mathbf{i} + y^3\mathbf{j} + z^3\mathbf{k}$ through the surface of the sphere $x^2 + y^2 + z^2 = R^2$.

$$A: \Phi = \frac{12\pi R^5}{5}$$

3. Compute the flow of the vector $\mathbf{r} = x\mathbf{i} + y\mathbf{j} + z\mathbf{k}$ through the exterior surface of the ellipsoid of equation $\frac{x^2}{a^2} + \frac{y^2}{b^2} + \frac{z^2}{c^2} = 1$.

$$A: \Phi = 4\pi abc$$

4. Compute the flow of the vector

$$\mathbf{V} = \left( z^2 - y^2 \right)\mathbf{i} + \left( x^2 - z^2 \right)\mathbf{j} + \left( y^2 - x^2 \right)\mathbf{k}$$

through the exterior surface of a closed bounded domain from $\mathfrak{R}^3$.

A: $\Phi = 0$

# REFERENCES

1. BÂRZĂ, I., *Analiză Matematică. Culegere de Probleme Rezolvate* (Mathematical Analysis. A collection of solved problems), Niculescu Printing House, Bucharest, 2002.

2. CIORĂNESCU, N., *Curs de Algebră şi Analiză Matematică* (Course of Algebra and Analysis), Ed. Tehnică, Bucharest, 1958.

3. CRAW, I., *Advanced Calculus and Analysis*, Univ. of Aberdeen, 2000.

4. COURANT, R., *Differential & Integral Calculus*, t.2, Blackie and Son Ltd, London and Glasgow, 1936.

5. DANKO, P.E., POPOV, A.G., *Vîsşaia Matematika v uprajneniah i zadachah*, Vîsşaia Şkola, Moscva, 1964 (Advances mathematics in problems and exercises).

6. NIŢĂ, A., STĂNĂŞILĂ, T., *1000 de probleme rezolvate şi exerciţii fundamentale pentru studenţi şi elevi* (1000 solved problems and fundamental exercices for students), Ed. BIC ALL, Bucharest, 1997.

7. PĂLTINEANU, G., *Analiză Matematică. Calcul integral*, (Mathematical Analysis. Integral Calculus) Ed. Agir, Bucureşti, 2004.

8. POPA, I., *Analiză Matematică. Calcul integral*, Matrix Rom, Bucureşti, 2000.

9. SOARE, M.V., TEODORESCU, P.P., TOMA, I., *Ordinary differential equations with applications to mechanics*, Springer, Dordrecht, 2006.

10. TOMA, I., *Analyse Mathématique. Calcul différentiel*, (Mathematical Analysis. Differential Calculus) Conspress, Bucharest, 2010.

11. TOMA, I., MOŞNEGUŢU, V., CONSTANTINESCU, Şt., *Analyse Mathématique. Équations différentielles ordinaires. Calcul intégral,* Ed. Conspress, Bucharest, 2014.

12. TOMA, I., MOSNEGUŢU, V., CONSTANTINESCU, ŞT., *Ordinary Differential Equations,* CreateSpace Independent Publishing Platform, 2016, ISBN - 13: 978-1540318015, ISBN - 10: 154031801X

Links:

13. GEORGE A. OSBORNE, S.B., *Differential and integral calculus, with exemples and applications,* Boston, U.S.A., D.C. Heath & CO., Publishers, 1906.
https://archive.org/stream/cu31924015990108#page/n3/mode/2up

14. BEATRIZ NAVARRO LAMEDA, NIKITA NIKOLAEV, *Integral Calculus,* Spring, 2016.
http://www.math.toronto.edu/nikolaev/files/MAT136_Lecture_Notes.pdf

15. DEEPAK BHARDWAJ, *Integral Calculus Made Easy,* Firewall Media, India, 2006
https://books.google.co.in/books?id=CD966TcWrQIC&pg=PA1&source=gbs_toc_r&hl=ro#v=onepage&q&f=false

16. CLYDE E LOVE, EARL DAVID RAINVILLE, *.Differential and integral calculus,* New York The Macmillan Company 1916.
https://archive.org/stream/differentialand00lovegoog#page/n20/mode/2up

17. JOSEPH EDWARDS, *A treatise on the integral calculus; with applications, examples and problems,* Macmillan and CO., 1921
https://archive.org/stream/treatiseonintegr01edwauoft#page/n27/mode/2up

These books can be also found online on the site

http://www.freebookcentre.net/Mathematics/Integral-Calculus-Books.h

18.   http://people.virginia.edu/~mah7cd/Math132/Ch15.pdf

19.   https://web.math.rochester.edu/people/faculty/edummit/docs/calc3_3_multiple_integration.pdf

20.   http://www.mathstat.concordia.ca/faculty/cdavid/EMAT212/solintegrals.pdf

21.   http://www.math.tamu.edu/courses/math601/book/Part2Chapter5.pdf

22.   http://www.math.toronto.edu/nikolaev/files/MAT136_Lecture_Notes.pdf

23.   http://bookboon.com/en/integration-and-differential-equations-ebook

24.   https://ocw.mit.edu/ans7870/resources/Strang/Edited/Calculus/Calculus.pdf

25.   http://library.umac.mo/ebooks/b31290735.pdf